Mastering Machine Learning Architecture and Solutions

From Design to Deployment

Mohammad Reza Mahdiani

Apress®

Mastering Machine Learning Architecture and Solutions: From Design to Deployment

Mohammad Reza Mahdiani
Calgary, AB, Canada

ISBN-13 (pbk): 979-8-8688-2526-2
ISBN-13 (electronic): 979-8-8688-2527-9
https://doi.org/10.1007/979-8-8688-2527-9

Managing Director, Apress Media LLC: Welmoed Spahr
Acquisitions Editor: Celestin Suresh John
Development Editor: Jim Markham
Coordinating Editor: Gryffin Winkler

Cover image designed by Freepik (www.freepik.com)

Distributed to the book trade worldwide by Springer Science+Business Media New York, 1 New York Plaza, New York, NY 10004. Phone 1-800-SPRINGER, fax (201) 348-4505, e-mail orders-ny@springer-sbm.com, or visit www.springeronline.com. Apress Media, LLC is a Delaware LLC and the sole member (owner) is Springer Science + Business Media Finance Inc (SSBM Finance Inc). SSBM Finance Inc is a **Delaware** corporation.

For information on translations, please e-mail booktranslations@springernature.com; for reprint, paperback, or audio rights, please e-mail bookpermissions@springernature.com.

Apress titles may be purchased in bulk for academic, corporate, or promotional use. eBook versions and licenses are also available for most titles. For more information, reference our Print and eBook Bulk Sales web page at http://www.apress.com/bulk-sales.

Any source code or other supplementary material referenced by the author in this book can be found here: https://www.apress.com/gp/services/source-code.

If disposing of this product, please recycle the paper

To my beloved family.

Table of Contents

About the Author

Mohammad Reza Mahdiani, Ph.D., is a technical leader and software architect with deep experience in large-scale AI systems and intelligent automation. His work bridges advanced research and industrial-grade execution, with a focus on engineering systems that deliver measurable and lasting business value.

He specializes in setting clear technical direction and designing production-ready solutions that emphasize architectural clarity, performance, and reliability at scale.

About the Technical Reviewers

Ramya Ravi specializes in technical marketing for AI/ML developer ecosystems and technical content creation. She has authored tutorials, guides, and demos that have helped thousands of Python and AI developers adopt and build with modern machine learning frameworks. Ramya has contributed to the PyTorch community through educational resources and developer engagement and has worked for industry leaders—Intel and Tata Consultancy Services (TCS). Her work focuses on making AI/ML technologies more accessible, practical, and impactful for developers worldwide.

Siddhant Agarwal is a seasoned Developer Relations professional with over a decade of experience building and scaling developer ecosystems globally. Currently leading Developer Relations across APAC at Neo4j and recognized as a Google Developer Expert in GenAI, he is passionate about empowering developers to reimagine data and AI through graph technology. Known for his unique ability to tell powerful stories through technology, Sid translates complex ideas into narratives that inspire action and innovation. With his signature "Local to Global" approach, he turns grassroots creativity into global impact. Previously at Google managing flagship developer programs, he continues to shape communities, share insights on GenAI and graph databases, and drive meaningful connections across the tech world. Learn more at meetsid.dev.

Introduction

In this book, we guide you in transforming abstract machine learning models into high-impact, robust, and scalable production systems. It is specifically designed for data scientists, ML engineers, and MLOps practitioners looking to master the entire lifecycle.

The main challenge in MLOps is bridging the gap between reliable production operation and model building. This book offers a structured approach to master these advanced skills across software engineering, specialized ML techniques, and DevOps.

Here, we cover the architecture of ML solutions across the end-to-end deployment lifecycle, including

- **Infrastructure and Deployment**: Mastering containerization (Docker) and orchestration (Kubernetes)
- **Quality Assurance**: Designing and implementing tests for data, features, and model performance
- **Model Management**: Techniques for versioning, automated retraining, and various drift handling
- **Responsible AI**: Ensuring model fairness, security, and transparency
- **Advanced Topics**: Exploring cutting-edge trends like agentic AI systems, transformers, and edge AI

To maximize learning, each chapter follows a specific approach including theory, practical code snippets, and a real-world step-by-step project. I recommend running all code samples by yourself for translating your theoretical knowledge into practical MLOps skills.

CHAPTER 1

Introduction to Machine Learning Architecture

Machine learning (ML) architecture refers to the layout and design principles for developing machine learning models. These rules include the development, deployment, and administration of models using machine learning. It comprises software and hardware components, including algorithms, data pipelines, and computational infrastructure necessary for training and model serving. Machine learning architecture is fundamentally designed to support all stages of workflow such as data analysis, preprocessing, model training, validation, and inference, guaranteeing a scalable, maintainable, and robust system.

Modern architectures usually use frameworks such as TensorFlow, PyTorch, or scikit-learn. In addition, they use distributed systems and cloud computing to manage big volumes of data and calculations. Choosing the appropriate model type is important in machine learning architecture. A suitable model selection guarantees reproducibility and supports batch or real-time inference. Knowing the structure of artificial intelligence is crucial for enhancing results, lowering expenses, and dealing with problems. These problems include data drift, model bias, and system transparency. In this chapter, we will introduce the core concepts of ML architecture, including key challenges and opportunities, and the overall machine learning lifecycle. In later chapters, we will dive deeper into these topics.

In this book, in each chapter or section, I try to include some case studies or personal experiences to help you learn more effectively. At the end of each chapter, we have a project to simulate a real-world case study about the content of the chapter. Are you ready? We can start our discussion with definitions.

M. R. Mahdiani, *Mastering Machine Learning Architecture and Solutions*,
https://doi.org/10.1007/979-8-8688-2527-9_1

What Is ML Architecture?

ML architecture provides the structural blueprint for creating an ML system that addresses a specific problem or task. It includes everything, such as dependencies, relationships, and processes. ML architecture is needed for effective creation, training, validation, distribution, and maintenance of machine learning models. It helps data scientists, engineers, and stakeholders see how various parts interact to achieve a desired result.[1] To get more details, let's see the components of the ML architecture.

Components of ML Architecture

Each fundamental component of ML architecture plays an essential role in the functioning of the system. The components usually are data ingestion, data preprocessing, feature engineering, model training, model validation, deployment, and monitoring.[2] The order presented above is typical in many practical projects; however, in some cases, teams may adjust the pipeline to be iterative. Figure 1-1 shows a basic summary of the fundamental elements of ML structure.

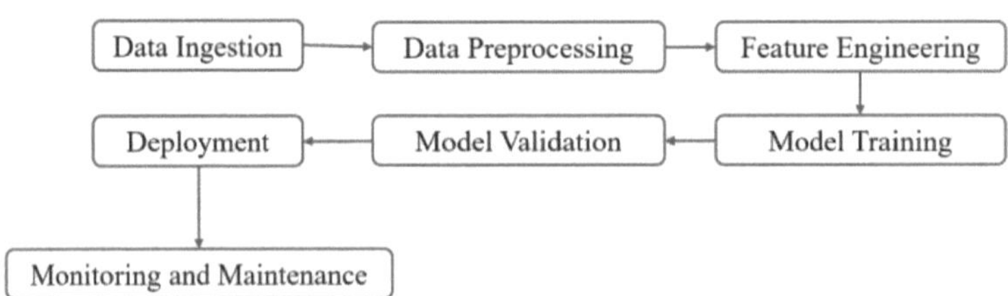

Figure 1-1. *Overview of a Typical Machine Learning Architecture*

These elements together form the machine learning life cycle. Later in this chapter, machine learning lifecycle will be discussed in more detail.

Machine learning (ML) has advanced significantly in recent years. Thus, choosing the right architecture is crucial in defining model performance, scalability, and interpretability. Architecture selection not only affects the learning process but also shapes how data is handled, how predictions are generated, and how systems are implemented in practical contexts.

In this section, we look at several machine learning approaches spanning from traditional methods such as supervised, unsupervised, and reinforcement learning to more sophisticated paradigms, including self-supervised learning, multi-agent systems, federated learning, neural architecture search (NAS), and edge AI. Figure 1-2 provides a visual classification of these ML approaches to assist viewers in understanding the relationships between conventional and modern techniques.

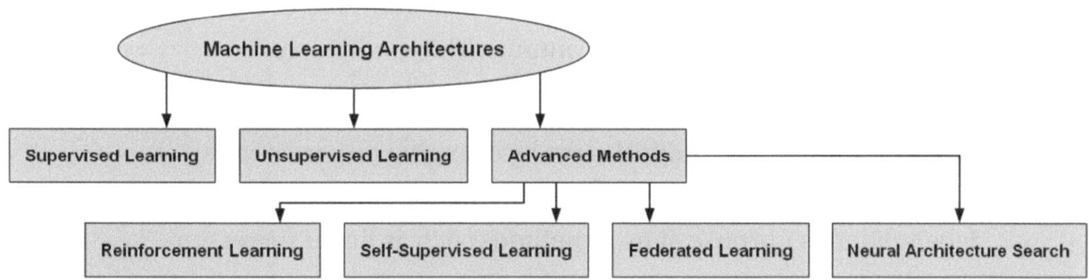

Figure 1-2. *Hierarchical Classification of Machine Learning Architectures*

Table 1-1 lists various machine learning approaches with a brief description of each. This table is a reference. We will discuss each architecture in more detail after this table.

Table 1-1. *Overview of Common Unsupervised Learning Algorithms*

Category	Description
Supervised Learning	Learning with labeled data (e.g., classification, regression)
Unsupervised Learning	Learning patterns from unlabeled data (e.g., clustering, PCA)
Reinforcement Learning	Learning through reward-based decision-making (e.g., robotics, games)
Self-Supervised Learning	Learning representations from data without explicit labels
Multi-agent Systems	Multi-agent interaction in dynamic environments
Federated Learning	Decentralized learning across multiple devices while preserving privacy
Neural Architecture Search (NAS)	Automated ML model discovery and optimization
Edge AI	Deploying ML models on edge devices for real-time processing

Supervised Learning

Many classic ML methods rely on supervised learning. Models in this paradigm are trained on a dataset in which each input corresponds to an output or label. The main goal is to learn a mapping from inputs to outputs that generalizes well to unseen data. For supervised learning, we have to consider the following steps:

1. **Preparation of Data**: The dataset includes feature vector inputs and corresponding label outputs. This could include normalization, data cleansing, and feature engineering.
2. **Model Training**: The model learns by changing its parameters to reduce the difference between its predictions and the true labels. For this purpose, it uses methods such as linear regression, decision trees, support vector machines, or artificial neural networks.
3. **Evaluation and Validation**: The model's performance is measured using metrics like accuracy, precision, recall, or mean squared error. Methods such as cross-validation help the model to generalize well.
4. **Deployment**: The model is used after training to predict new data that usually gets incorporated into live applications.

Supervised learning is commonly used in several areas, from financial predictions and image and speech recognition to medical diagnostics. Common supervised learning methods, their fundamental properties, and typical applications are summarized in Table 1-2.

Table 1-2. *Summary of Common Supervised Learning Algorithms*

Algorithm	Key Characteristics	Applications
Linear Regression	Assumes linear relationship; interpretable	House price prediction, trend analysis
Logistic Regression	Models binary or multi-class problems; probabilistic outputs	Spam detection, disease classification
Decision Trees	Nonlinear decision boundaries; easy to interpret	Customer segmentation, risk assessment
Support Vector Machines (SVM)	Effective in high-dimensional spaces; margin maximization	Image classification, handwriting recognition
Neural Networks	Complex feature extraction, high computational cost	Speech recognition, natural language processing

Now, let's have an example code for supervised learning. The code below uses a decision tree classifier on the well-known Iris dataset. This sample illustrates the entire workflow from data loading to model evaluation.

Listing 1-1. Decision Tree Classifier

```
from sklearn.datasets import load_iris
from sklearn.tree import DecisionTreeClassifier
from sklearn.model_selection import train_test_split
from sklearn.metrics import accuracy_score

# Load the Iris dataset
iris = load_iris()

# Split into training and testing sets
X_train, X_test, y_train, y_test = train_test_split(
    iris.data, iris.target, test_size=0.2, random_state=42
)

# Initialize and train the classifier
clf = DecisionTreeClassifier(random_state=42)
clf.fit(X_train, y_train)
```

```
# Make predictions and evaluate accuracy
y_pred = clf.predict(X_test)
print("Model Accuracy:", accuracy_score(y_test, y_pred))
```

Expected Results

```
Model Accuracy: 1.0
```

This code demonstrates the practical application of a decision tree in a supervised learning context, highlighting its simplicity and effectiveness for classification tasks. It should be mentioned that for small classification problems, accuracy 1 is not that weird; however, if we want to predict a value, accuracy 1 is too good to be true. As you saw, supervised learning is very straightforward; you provide a dataset containing features (inputs) and labels (outputs), and the algorithm learns a mapping from inputs to outputs. In the next parts, we will talk about more complicated architectures.

Unsupervised Learning

Unsupervised learning is a group of algorithms that work on unlabeled data to discover unknown patterns within the dataset. Unsupervised learning has several fundamental categories:

Clustering: K-Means, DBSCAN, and hierarchical clustering algorithms group data points based on similarity metrics. The purpose is to find naturally occurring groups in the data.

Dimensionality Reduction: Methods such as principal component analysis (PCA) and t-SNE (t-Distributed Stochastic Neighbor Embedding) reduce the number of features while preserving as much relevant information (variance) as possible. It helps to visualize and eliminate noise.

Representation Learning: Techniques such as autoencoders and manifold learning learn compact or informative representations of data. These representations can serve as preprocessing for other ML tasks.

The following code shows an example of unsupervised learning. This code demonstrates K-Means clustering on a small synthetic dataset. The algorithm partitions the data into clusters based on Euclidean distance.

Listing 1-2. K-Means Clustering

```
#K-Means Clustering Example
from sklearn.cluster import KMeans
import numpy as np

# Sample data
X = np.array([[1, 1], [1.5, 1.8], [5, 8], [8, 8], [1, 0.6], [9, 11]])

# Create and train the K-Means model (e.g., 2 clusters)
kmeans = KMeans(n_clusters=2, random_state=0)
kmeans.fit(X)

# Get cluster labels for each data point
labels = kmeans.labels_

# Print cluster assignments
print(f"Cluster Labels: {labels}")

# Get the coordinates of the cluster centers
centroids = kmeans.cluster_centers_
print(f"Centroids: {centroids}")

# Predict the cluster for new data points
new_data = np.array([[2, 2], [7, 9]])
predictions = kmeans.predict(new_data)
print(f"Predictions for new data: {predictions}")
```

Expected Results

```
Cluster Labels: [1 1 0 0 1 0]
Centroids: [[7.33333333 9.        ]
 [1.16666667 1.13333333]]
Predictions for new data: [1 0]
```

Before talking about the above code, read it again and try to use unsupervised learning in a practical problem. The above code illustrates how K-Means clustering can reveal natural groupings within data, which is invaluable for exploratory data analysis and unsupervised pattern recognition. As we proceed, learning models are getting more exciting.

Reinforcement Learning

Reinforcement learning (RL) is a machine learning paradigm in which an agent learns to make decisions by interacting with an environment. At each step, the agent takes an action, the environment responds with a new state and a reward signal, and the agent uses this feedback to improve its behavior. Reinforcement learning considers the following parts:

Interaction Between Agent and Environment: The agent operates in the environment and gets feedback as rewards or penalties.

Policy Learning: The agent's objective is to create a policy. A policy is a mapping from states to actions. It simultaneously increases the expected cumulative rewards over time.

Exploration vs. Exploitation: Reinforcement learning consists of two trade-offs for finding the most optimal actions. Exploration involves trying fresh behaviors to see their effects, and exploitation is using well-known behaviors that give good rewards.

Techniques: Q-learning, policy gradients, and actor-critic methods are among the most common techniques that are used to maximize the performance of the agent.

Reinforcement learning is commonly used in cases that need sequential decision-making. These include autonomous vehicles, robotics, and gaming. It helps robots to learn difficult tasks via trial and error. To give you a more practical case, consider a model that plays chess with users. To train the model, it can play chess with itself and learning which ways lead to success and which lead to failure. The sample code below shows an example of reinforcement learning.

Listing 1-3. A Basic Q-Learning Update

```
import numpy as np
q_table = np.zeros((5, 5))

# Define an example state, action, and reward
state = 2
action = 1
reward = 10
learning_rate = 0.1
discount_factor = 0.9

# Simplified Q-learning update rule:
```

```
# Q(s, a) = Q(s, a) + learning_rate * [reward + discount_factor * 
max(Q(s') - Q(s, a))]
q_table[state, action] = q_table[state, action] + learning_rate * (reward + 
discount_factor * np.max(q_table[state]) - q_table[state, action])

print("Updated Q-Table:", q_table)
```

Expected Results

```
Updated Q-Table: [[ 0.   0.   0.   0.   0. ]
                  [ 0.   0.   0.   0.   0. ]
                  [ 0.   1.   0.   0.   0. ]
                  [ 0.   0.   0.   0.   0. ]
                  [ 0.   0.   0.   0.   0. ]]
```

The below code is another example of reinforcement learning that shows how RF learning can be used in chess.

Listing 1-4. Simplified Code for Using Reinforcement Learning for Chess

```
import chess
import chess.engine
import random
import numpy as np
import pickle

# Simple Q-table as a dictionary
q_table = {}

# Parameters
alpha = 0.1      # Learning rate
gamma = 0.9      # Discount factor
epsilon = 0.2    # Exploration rate
episodes = 1000

# Function to convert board to a hashable state
def board_to_state(board):
    return board.fen()
```

```
# Choose action with epsilon-greedy
def choose_action(board, legal_moves):
    if random.random() < epsilon:
        return random.choice(legal_moves)
    state = board_to_state(board)
    if state not in q_table:
        return random.choice(legal_moves)
    move_scores = q_table[state]
    return max(move_scores, key=move_scores.get)

# Reward function
def get_reward(board):
    if board.is_checkmate():
        return 1
    elif board.is_stalemate() or board.is_insufficient_material():
        return 0.5
    elif board.is_game_over():
        return -1
    return 0

# Training loop
for episode in range(episodes):
    board = chess.Board()
    while not board.is_game_over():
        state = board_to_state(board)
        legal_moves = list(board.legal_moves)

        if state not in q_table:
            q_table[state] = {move: 0 for move in legal_moves}

        action = choose_action(board, legal_moves)
        board.push(action)

        reward = get_reward(board)
        next_state = board_to_state(board)
        if next_state not in q_table:
            q_table[next_state] = {move: 0 for move in board.legal_moves}
```

```
        if not board.is_game_over():
            next_max = max(q_table[next_state].values()) if q_table
            [next_state] else 0
        else:
            next_max = 0

        # Q-learning update
        q_table[state][action] += alpha * (reward + gamma * next_max -
        q_table[state][action])

    if episode % 100 == 0:
        print(f"Episode {episode} completed.")

# Save Q-table
with open("q_chess_agent.pkl", "wb") as f:
    pickle.dump(q_table, f)
```

Expected Results

```
Episode 0 completed.
Episode 100 completed.
Episode 200 completed.
Episode 300 completed.
Episode 400 completed.
Episode 500 completed.
Episode 600 completed.
Episode 700 completed.
Episode 800 completed.
Episode 900 completed.
```

You should see results similar to the above. I suggest reading the above code line by line and thinking about how it works.

Self-Supervised Learning

Self-supervised learning (SSL) is a major machine learning paradigm in which models learn directly from the structure of unlabeled data. SSL generates its own training signals by defining tasks that can be solved using only the input data instead of relying on manually labeled datasets. This makes SSL especially valuable when labeled data is limited or expensive to obtain. SSL is used especially if labeling data is expensive. In this case, this strategy is very useful. SSL normally includes

Pretext Tasks: The model is trained on pretext tasks (e.g., generating a masked part of an image or sentence) that don't need manual labeling.

Representation Learning: The model learns valuable data representations by addressing these pretext tasks that can be adjusted for downstream uses.

In natural language processing (e.g., BERT, GPT) and computer vision, SSL has been very effective in transferability, allowing models to achieve state-of-the-art performance with little labeled data.

SSL is widely used in language models that are trained on a large text corpus. The code below shows how to compute cosine similarity between two embedding vectors which is commonly done when evaluating representations produced by an SSL-trained model.

Listing 1-5. Contrastive Learning Similarity Calculation

```
import torch
import torch.nn.functional as F

# Define sample embedding vectors
embedding1 = torch.tensor([0.1, 0.2, 0.3])
embedding2 = torch.tensor([0.1, 0.2, 0.35])

# Compute cosine similarity
similarity = F.cosine_similarity(embedding1.unsqueeze(0), embedding2.
unsqueeze(0))
print("Similarity Score:", similarity.item())
```

Expected Results

```
Similarity Score: 0.998
```

This example demonstrates how self-supervised learning techniques can use similarity measures to refine feature representations without requiring labeled data. Are you ready for a more complicated example of self learning? The code below shows a more complicated example of a self-supervised learning. An explanation about each step is provided as a comment in the code.

Listing 1-6. Self-Learning Sample Code

```
import torch
import torch.nn as nn
import torch.nn.functional as F
from torchvision import transforms, datasets, models
from torch.utils.data import DataLoader

# Data augmentation for SSL
transform = transforms.Compose([
    transforms.RandomResizedCrop(32),
    transforms.RandomHorizontalFlip(),
    transforms.ColorJitter(0.4, 0.4, 0.4, 0.1),
    transforms.ToTensor()
])

dataset = datasets.CIFAR10(root="./data", train=True, download=True,
transform=transform)
loader = DataLoader(dataset, batch_size=256, shuffle=True, drop_last=True)

# Contrastive projection head
class ProjectionHead(nn.Module):
    def __init__(self, in_dim):
        super().__init__()
        self.fc1 = nn.Linear(in_dim, 256)
        self.fc2 = nn.Linear(256, 128)

    def forward(self, x):
        x = F.relu(self.fc1(x))
        return self.fc2(x)

# SimCLR Model (backbone + head)
class SimCLR(nn.Module):
    def __init__(self):
        super().__init__()
        backbone = models.resnet18(weights=None)
        backbone.fc = nn.Identity()
        self.encoder = backbone
        self.projector = ProjectionHead(512)
```

```
    def forward(self, x):
        h = self.encoder(x)
        z = self.projector(h)
        return F.normalize(z, dim=1)

# NT-Xent Loss (Contrastive)
def nt_xent_loss(z1, z2, temperature=0.5):
    batch_size = z1.size(0)
    z = torch.cat([z1, z2], dim=0)
    similarity = torch.mm(z, z.t())

    labels = torch.arange(batch_size).repeat(2)
    labels = labels.to(z.device)

    mask = torch.eye(2 * batch_size).bool().to(z.device)
    similarity = similarity[~mask].view(2 * batch_size, -1)

    return F.cross_entropy(similarity / temperature, labels)


# Training Loop
device = "cuda" if torch.cuda.is_available() else "cpu"
model = SimCLR().to(device)
optimizer = torch.optim.Adam(model.parameters(), lr=1e-3)

for epoch in range(5):
    for imgs, _ in loader:
        imgs1 = imgs.to(device)
        imgs2 = imgs.to(device)

        z1 = model(imgs1)
        z2 = model(imgs2)

        loss = nt_xent_loss(z1, z2)
        optimizer.zero_grad()
        loss.backward()
        optimizer.step()

    print(f"Epoch {epoch}, Loss: {loss.item():.4f}")
```

Expected Results

```
Downloading https://www.cs.toronto.edu/~kriz/cifar-10-python.tar.gz to ./
data\cifar-10-python.tar.gz
100%|████████████████████████████████████████████████████████████████| 170M/170M [00:05<00:00,
30.5MB/s]
Extracting ./data\cifar-10-python.tar.gz to ./data
Epoch 0, Loss: 5.3261
Epoch 1, Loss: 5.3036
Epoch 2, Loss: 5.3019
Epoch 3, Loss: 5.2896
Epoch 4, Loss: 5.2942
```

In the above code, we used one of the most common techniques in self-supervised learning which is called SimCLR-style contrastive learning system.

Multi-agent Systems

Multi-agent Systems (MAS) consist of several independent agents interacting with each other to address sophisticated issues. In these systems, every agent functions according to local knowledge but in coordination with communication and common goals. Some of the essential features of multi-agent systems are the following:

Decentralized Decision-Making: Every agent acts independently, which can cause emergent behavior when agents interact.

Communication: Agents communicate to coordinate their activities and prevent clashes, as in communication protocols.

Coordination and Cooperation: The system is set up for agents to collaborate on reaching an international goal, therefore usually creating strong and flexible solutions.

Adaptability: For rapidly changing systems where centralized control is not possible, MAS is extremely adjustable.

MAS can be effectively used in robotics that covers swarms of drones or independent robots executing search and rescue, traffic control that adaptive traffic lights react in real time, and collaborative surveillance and management found in Internet of Things environments, in distributed sensor networks. The following is a sample Python class showing how simple communication among agents in a multi-agent system works.

Listing 1-7. Agent Communication Simulation

```
class Agent:
    def __init__(self, name):
        self.name = name

    def communicate(self, message):
        print(f"{self.name} received message: {message}")

# Simulate communication between agents
agent_a = Agent("Agent A")
agent_b = Agent("Agent B")

agent_a.communicate("Initiate cooperative behavior.")
agent_b.communicate("Acknowledged. Executing coordinated action.")
```

This code (or pseudo code!) simulates how agents in a multi-agent system may exchange information to improve their actions, a crucial component for achieving system-wide objectives. It just simulates the communication between multiple agents and is not a real technical code.

Neural Architecture Search (NAS)

Neural architecture search automates the design of neural network architecture. It significantly reduces the need for manual intervention from architecture design experts. This encompasses searching through a predefined space of architecture and, iteratively, optimizing performance metrics such as accuracy and latency. NAS involves

Search Space Definition: Identifying the set of possible architectures or variations in layer types, connections, and hyperparameters.

Optimization Strategy: Using reinforcement learning, evolutionary algorithms, or gradient-based techniques to assess and enhance candidate architectures.

Performance Evaluation: Every architecture is trained (either fully or partially) and evaluated so that the best-performing designs can be selected.

Resource Efficiency: Except for the previously mentioned aspects of design, NAS is ideally optimized in such a way that it performs to specification and minimizes resource consumption.

NAS can be used effectively in automated model design. With its high speed, it gives a new approach to automation in computer vision, natural language processing, and beyond. Additionally, NAS is suitable for cases that require continuous adaptation of architectures based on data and task requirements. In the following, an example code of how to implement NAS is shown.

Listing 1-8. Neural Architecture Search with AutoKeras

```
import random

# Define a simple search space (you'd have a much larger one in reality)
operations = ["conv3x3", "conv5x5", "maxpool", "avgpool"]
num_layers = 4

def generate_architecture():
    """Generates a random neural network architecture."""
    architecture = []
    for _ in range(num_layers):
        op = random.choice(operations)
        architecture.append(op)
    return architecture

def evaluate_architecture(architecture):
    """Simulates evaluating an architecture (in reality, you'd
    train it)."""
    # This is a placeholder - replace with actual training and validation
    # For demonstration, assign a random "accuracy" between 0 and 1.
    accuracy = random.uniform(0, 1)
    return accuracy

def simple_nas():
    """Performs a very basic Neural Architecture Search (random search)."""
    num_architectures_to_try = 10  # Try 10 random architectures

    best_architecture = None
    best_accuracy = 0

    for _ in range(num_architectures_to_try):
        architecture = generate_architecture()
```

```
        accuracy = evaluate_architecture(architecture)

        print(f"Architecture: {architecture}, Accuracy: {accuracy:.4f}")

        if accuracy > best_accuracy:
            best_accuracy = accuracy
            best_architecture = architecture

    print(f"\nBest Architecture: {best_architecture}, Best Accuracy: {best_
    accuracy:.4f}")

# Run the NAS
simple_nas()
```

Expected Results

```
Architecture: ['maxpool', 'maxpool', 'avgpool', 'avgpool'], Accuracy: 0.0339
Architecture: ['conv3x3', 'maxpool', 'conv5x5', 'maxpool'], Accuracy: 0.9647
Architecture: ['avgpool', 'conv3x3', 'conv3x3', 'conv3x3'], Accuracy: 0.3749
Architecture: ['conv5x5', 'avgpool', 'avgpool', 'conv5x5'], Accuracy: 0.3724
Architecture: ['maxpool', 'conv5x5', 'conv3x3', 'conv5x5'], Accuracy: 0.4574
Architecture: ['conv3x3', 'conv5x5', 'conv5x5', 'conv3x3'], Accuracy: 0.6810
Architecture: ['conv3x3', 'avgpool', 'conv5x5', 'conv5x5'], Accuracy: 0.7438
Architecture: ['conv5x5', 'avgpool', 'conv5x5', 'conv3x3'], Accuracy: 0.3561
Architecture: ['conv5x5', 'maxpool', 'conv3x3', 'conv3x3'], Accuracy: 0.1684
Architecture: ['conv5x5', 'conv3x3', 'avgpool', 'maxpool'], Accuracy: 0.4126

Best Architecture: ['conv3x3', 'maxpool', 'conv5x5', 'maxpool'], Best
Accuracy: 0.9647
```

In this section of this chapter, we briefly talked about various ML approaches from traditional to modern ones.

Layered Architecture in Machine Learning

The layered architecture concept is widely used in the design of machine learning systems. This entails clustering similar processes into logical layers. A layered architecture allows for better organization, scaling, and maintenance of the ML system. A standard model can be classified under the following three general layers:

Data Layer: As the foundational layer, the data layer is the cornerstone of any ML system that comprises the handling of data. The data layer considers various aspects of data collection, storage, transformation, and retrieval. This layer could include data pipelines that preprocess raw data to a more useful format. The data layer changes data in a way that databases or data lakes can support the quality and consistency of it. Data management is important because much will rely on its functioning; after all, the performance of the ML model could be greatly affected by the very nature of the workable data and its accessibility.

Model Layer: The model layer focuses on core machine learning procedures, like development, training, optimization, and versioning of models. The model layer involves choosing the best algorithms, working on hyperparameters, training with datasets, and validating for accuracy and fairness. Also, maintaining a repository for effectively managing various model versions is a part of the model layer. Additionally, keeping efficiency in deploying the latest versions may also be in this layer.

Service Layer: The service layer connects the ML system to the front-end application and other internal systems; it deals with deploying trained models and exposing these models through APIs. The service layer enables predictions or inferences made by the models to be consumed by larger systems, such as web applications, mobile apps, enterprise applications, or any other consumer. It goes further in monitoring model performance in production, diagnosing scaling issues, and ensuring system stability and reliability.

Table 1-3 summarizes the layered design principles as presented in this section. Technologies and tools are also described in this table to support further analysis.

Table 1-3. *Overview of ML Architecture Layers*

Layer	Description	Components Included	Technologies/Tools
Data Layer	Manages data storage, transformation, etc.	Data Ingestion, Data Preprocessing	Hadoop, Spark, SQL, Kafka
Model Layer	Responsible for training and optimizing models.	Feature Engineering, Model Training, Model Validation	scikit-learn, TensorFlow, PyTorch, Keras
Service Layer	Exposes trained models for consumption through APIs.	Deployment, Monitoring, Maintenance	Flask, FastAPI, Kubernetes, Docker, Grafana

Importance of ML Architecture

Machine learning architecture is the central part for making ML solutions efficient, maintainable, and scalable. The proper architecture describes the way from data acquisition to making the final prediction. Architecture helps in making decisions to keep all the components working well and aligned with each other.

As an example, building a recommendation system for an ecommerce platform, the architecture includes the source of data (user interactions, product descriptions, etc.), feature engineering (user behavior features, product similarity), and where to deploy the models (real-time APIs). Also, architecture determines the effectiveness of the solution and its scalability as data grows.

Challenges and Opportunities

There are quite a few challenges associated with architecting an ML system. We also discuss the opportunities, innovation, and improvement to deal with those challenges.

Scalability Challenges

ML systems are normally built on much smaller subsets of data and scaled to much larger datasets in production. That is to say, scalability implies generalizing an architecture to address data growth, providing wider computational support, and allowing for distributed processing needs. Scalability is one of the main challenges in ML projects.

Opportunity for Scalability

The problem of scalability can be overcome by cloud-native architectures and distributed systems, which enable elasticity in data handling and computation. Services like AWS S3 provide scalable storage, and Google BigQuery enables large-scale data processing. Together, these cloud services support scalable ML pipelines, allowing inference workloads to automatically adjust to demand, taking away the load from the underlying infrastructure.

Latency and Real-Time Processing

Another big challenge is latency. This is more important, especially in applications that need real-time predictions, such as recommendation systems in ecommerce websites or fraud detection systems in financial applications. These cases need near-instant results for a seamless user experience, and high latency could degrade the user experience and affect the model's overall effectiveness.

Opportunity for Real-Time Systems

Concerning latency, it is quite reasonable to consider edge computing and optimized inference engines. Edge devices bring computation closer to the data source, resulting in less response time. In addition, lightweight models and frameworks such as TensorFlow Lite or ONNX Runtime make it possible to accommodate real-time processing for latency-sensitive applications. However, deploying large, compute-intensive models on edge hardware is challenging. We will have an extensive discussion about edge computing in Chapter 10. Model optimizations are another great way to reduce the load while maintaining the accuracy. If the above solutions do not work, the other option is to use additional GPU instances which can further reduce inference time.

Model Versioning and Reproducibility

The management of model versions and tracking of reproducible results in diverse environments is difficult. As environments, dependencies, and datasets evolve, retraining models makes it harder to track the model changes, data used, and how performance metrics shifted over time.

Opportunity in Model Versioning

MLOps practices and tools such as MLflow, DVC, and Kubeflow help address the above challenges by enabling structured model, data, and pipeline versioning. These systems make it easier to track experiments, maintain reproducibility, and manage the full ML lifecycle.

Operational Complexity

Deploying ML models into production introduces significant operational complexity. Complexity usually arises from the existence of various teams working on the same project. In addition, misalignment between data engineering and data science workflows can be another source of operational complexity. Additionally, the need to operate infrastructure and dependencies can result in more complicated situations.

Opportunity to Adopt Automation

Automation is essential for reducing operational complexity. CI/CD pipelines adapted for machine learning will help automate training, validation, deployment, and monitoring steps. MLOps practices support these automated workflows, reducing manual intervention and ensuring a consistent, reliable deployment cycle.

Data Quality and Bias

The challenge for ML systems includes data quality and bias. Poor-quality data leads to under-performing models, and biased input leads to predictions considered unfair or discriminatory. Mitigating these challenges is doable by data validation, handling missing or inconsistent data, and applying bias mitigation techniques.

There is an opportunity to establish fairness and transparency in machine learning systems. With the help of ethical AI practices and fairness libraries like IBM AI Fairness 360, practitioners may detect and mitigate biases in data and models. This improves trust from users and ensures compliance with regulations.

Summary of Challenges and Opportunities

Table 1-4 provides a summary of the key challenges in ML architecture and the corresponding opportunities to address them.

***Table 1-4.** Challenges and Opportunities in ML Architecture*

Challenge	Description	Opportunity
Scalability	Handling data growth and distributed processing	Cloud-native architectures, distributed systems
Latency	Reducing the time taken to generate predictions	Edge computing, optimized inference engines
Versioning and Reproducibility	Managing versions and reproducing results across environments	MLOps practices, tools like MLflow and DVC
Operational Complexity	Moving from development to production with minimal friction	CI/CD pipelines, automation tools
Data Quality and Bias	Ensuring data quality and minimizing bias	Fairness libraries, ethical AI practices
Explainability	Making ML models interpretable and understandable for stakeholders	Tools like SHAP, LIME, and explainable AI frameworks
Security	Protecting ML models and data against adversarial attacks and unauthorized access	Robust encryption, adversarial training, secure model deployment
Resource Efficiency	Managing computational and energy costs associated with training and inference	Efficient algorithms, model compression techniques like pruning and quantization
Integration with Legacy Systems	Incorporating ML systems with existing enterprise software and workflows	APIs, middleware solutions, and flexible microservice-based architectures
Continuous Learning	Adapting to new data and evolving patterns	Online learning frameworks, reinforcement learning approaches
Ethical Considerations	Ensuring compliance with privacy laws, ethical guidelines, and societal norms	Privacy-preserving techniques like differential privacy and federated learning
Diversity of Use Cases	Designing ML systems for varying applications across industries	Modular and configurable ML architectures
Monitoring and Maintenance	Ensuring long-term model performance and detecting drift	Continuous monitoring tools, model retraining pipelines, and automated alerts for performance decay

Overview of ML Lifecycle

The ML lifecycle starts with understanding the problem and ends with deploying the model and monitoring its effectiveness. Familiarity with the ML lifecycle is crucial for managing operational environments alongside the development and scaling of ML solutions. Moreover, the ML lifecycle offers a systematic way of building ML systems, containing several iterative stages.[4]

Stages of the ML Lifecycle

The ML lifecycle consists of the following stages:

Defining the Problem and Gathering the Requirements: This phase emphasizes the need to deeply state the business problem and to describe the success criteria that the ML solution should meet. It includes pertinent data, business goal setting, and estimating what impact the solution can have.

Data Collection and Understanding: In this step, data is collected from various locations like repositories, APIs, or even from some vendors. Data understanding occurs alongside collection and involves exploring relationships, correlations, distributions, and anomalies that may influence model design.

Data Preprocessing and Feature Engineering: After collection, there is a need to encode the data appropriately, and the data needs to be cleaned through preprocessing. This includes deleting duplicate entries, filling out available missing values, normalizing features, addressing inconsistencies, and transforming categorical variables into numerical formats. Feature engineering extracts meaningful attributes from raw data to improve model performance.

Model Development: Prepared data is used to train models in machine learning. In the course of solving the problem, different algorithms and approaches might be tried out to see which one solves the problem more effectively. Also, model performance tuning is done with the help of hyperparameters.

Model Assessment: Finally, after the training phase has been completed, models will be evaluated with new, unseen data. With metrics like accuracy, precision, recall, F1 score, and AUC-ROC, we evaluate the predictive capability of a model for scoring as well as detecting anything that possibly went wrong, for example, overfitting.

Model Deployment: Once the model is validated, it gets deployed to production. The models can serve predictions live or on-the-fly. This is the deployment of the model, running the model on a server, exercising and calling using REST APIs, or deploying on Edge devices too.

Model Monitoring and Maintenance: Once deployed, monitor the model performance in practice and monitor for degradation of performance/health. Monitoring is for checking prediction accuracies, detecting concept drift, or emerging biases from the model. To keep the model efficient all the time, we need to re-train the model on new data periodically.

Iterative Nature of the ML Lifecycle

ML lifecycle is an iterative process. Usually, model performance is suboptimal and needs multiple rounds of data collection, preprocessing, feature engineering, and retraining to lead to an optimal model.[4] Figure 1-3 represents the iterative nature of the ML lifecycle using a radar chart. This figure shows that the ML lifecycle has multiple stages, ranging from problem definition, data collection, preprocessing, and feature engineering up to model evaluation, deployment, and monitoring. The chart provides a nice visual of the ML incremental model refinement process and how tweaking one stage almost always revisits the previous model.

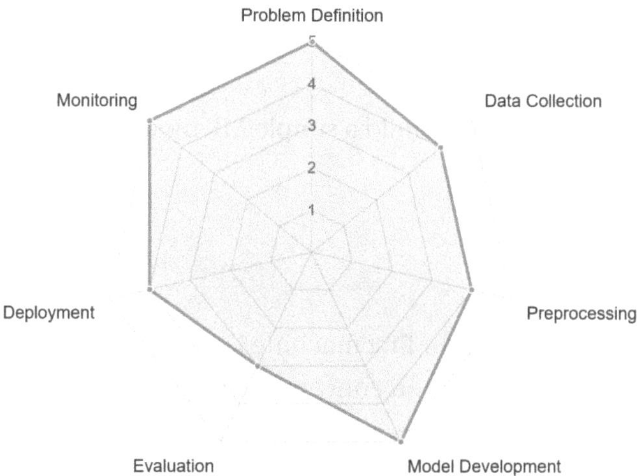

Figure 1-3. *Iterative Nature of the Machine Learning Lifecycle*

All of these stages are intertwined, meaning that one stage can often require going back to prior stages to improve it. For example, issues identified during model evaluation may reveal problems in feature engineering, or the model is relying only on limited data sources.

Importance of Understanding the ML Lifecycle

Why should you care about ML lifecycle? As a prerequisite, there are multiple reasons why the understanding of the ML lifecycle is fundamental. Here we briefly discuss them.

Project Management: Using a structured lifecycle, the teams can give better shape to their ML projects, accelerate progress, and avoid common obstacles.

Better Collaboration for Teams: Different teams, data engineers, data scientists, and ML engineers can collaborate more effectively.

Continuous Improvement: The lifecycle is iterative; therefore, models are continuously enhanced to learn from new data and business dynamics. This results in good models with improved performance over time.

Now, let's take a look at our project for this chapter and review the learnings in a more practical way.

Project: Building a Simple Machine Learning Pipeline

In this section, we will show how to build a simple ML model. Try to implement and execute the project by yourself.

Project Objectives

In this project, you will develop your first machine learning (ML) pipeline to help you get a firm grasp of all the key elements and concepts of an ML stack. It will teach you how your code should be broken into layers and tackle the usual gotchas around data quality, scalability, and operational complexity. Reading through this tutorial, you will master the visualization of the ML pipeline and develop a solid implementation for a basic classification use case.

Project Description

This end-to-end ML pipeline includes in-depth coverage of almost all tasks, from the data you read, through the training, evaluating, and deploying your model, as well as some visualizations where you can preview/adjust the model's prediction. At the end of this project, we will develop and run our working classifier on the Iris dataset, which is a canonical machine learning benchmark. The data contains 150 samples of iris flowers with four features (sepal length, sepal width, petal length, and petal width).

The project is structured into eight steps. We begin by importing the required libraries and loading the Iris dataset. We then explore and describe the data. Next, we perform preprocessing tasks such as handling missing values, encoding categorical variables, splitting the dataset into training and test sets, and applying feature scaling. After preprocessing, you will train a Random Forest classifier and evaluate its performance using multiple metrics and visualization tools. Finally, we will simulate model deployment using a prediction function and analyze feature importances that explain the model's decisions.

Step 1: Importing Libraries, We Need

First, import all libraries required for data manipulation, visualization, and machine learning. We want to prepare our environment with different tools, like `pandas` and `NumPy` for data manipulation, `matplotlib/seaborn` for visualization (although this could be handled differently), and `scikit-learn` for the models and evaluation. The code below imports the required libraries.

Listing 1-9. Step 1: Importing Libraries

```
import pandas as pd
import numpy as np
import matplotlib.pyplot as plt
import seaborn as sns

# Scikit-learn libraries for model building and evaluation
from sklearn.model_selection import train_test_split
from sklearn.preprocessing import StandardScaler
```

```
from sklearn.ensemble import RandomForestClassifier
from sklearn.metrics import classification_report, confusion_matrix,
accuracy_score

# Suppress warnings for cleaner output
import warnings
warnings.filterwarnings('ignore')
```

Explanation

In the code snippets, there are comments that explain more about the code.

Step 2: Data Ingestion

In this step, we load the Iris dataset and perform a basic analysis of the data. We import the dataset using a scikit-learn function and convert it into a NumPy array. This step is important because it prepares the data for further processing and ensures it is in a format suitable for the following stages of the pipeline. The code below demonstrates the data ingestion process.

Listing 1-10. Step 2: Data Ingestion

```
from sklearn.datasets import load_iris

# Load Iris dataset
iris = load_iris()
# Create a DataFrame
df = pd.DataFrame(data=iris.data, columns=iris.feature_names)
df['species'] = iris.target
# Map numerical target to actual species names
df['species'] = df['species'].map({0: 'setosa', 1: 'versicolor', 2:
'virginica'})

# Display the first five rows of the dataset
df.head()
```

Expected Result

sepal length (cm)	sepal width (cm)	petal length (cm)	petal width (cm)	species	
0	5.1	3.5	1.4	0.2	setosa
1	4.9	3.0	1.4	0.2	setosa
2	4.7	3.2	1.3	0.2	setosa
3	4.6	3.1	1.5	0.2	setosa
4	5.0	3.6	1.4	0.2	setosa

Step 3: Data Exploration and Visualization

The goal of this step is to cover the statistical nature and distribution of features in the dataset. You will create summary statistics and plots to understand the dataset. Look for patterns in features and verify preprocessing needs. Data processing will be comprehensively discussed in the next chapter. Here, as we want to have a simple project that includes all steps of the ML lifecycle, we will include the data cleaning. The code below explores and visualizes the data.

Listing 1-11. Step 3: Data Exploration and Visualization

```
# Summary statistics of the dataset
df.describe()
```

Expected Result

	sepal length (cm)	sepal width (cm)	petal length (cm)	petal width (cm)
count	150.000000	150.000000	150.000000	150.000000
mean	5.843333	3.057333	3.758000	1.199333
std	0.828066	0.435866	1.765298	0.762238
min	4.300000	2.000000	1.000000	0.100000
25%	5.100000	2.800000	1.600000	0.300000
50%	5.800000	3.000000	4.350000	1.300000
75%	6.400000	3.300000	5.100000	1.800000
max	7.900000	4.400000	6.900000	2.500000

The code below is for generating some visualizations for better understanding the data. With this code, we get 2×2 grids of histograms (one for each feature), which is illustrated in Figure 1-4. So that you can visually check the distribution and density of the data. KDE overlay gives you a better visual feel of data distribution.

Listing 1-12. Visualizing the Distribution of Each Feature

```
plt.figure(figsize=(12, 8))
for idx, column in enumerate(df.columns[:-1]):
    plt.subplot(2, 2, idx+1)
    sns.histplot(data=df, x=column, kde=True, bins=20)
    plt.title(f'Distribution of {column}')
plt.tight_layout()
plt.show()
```

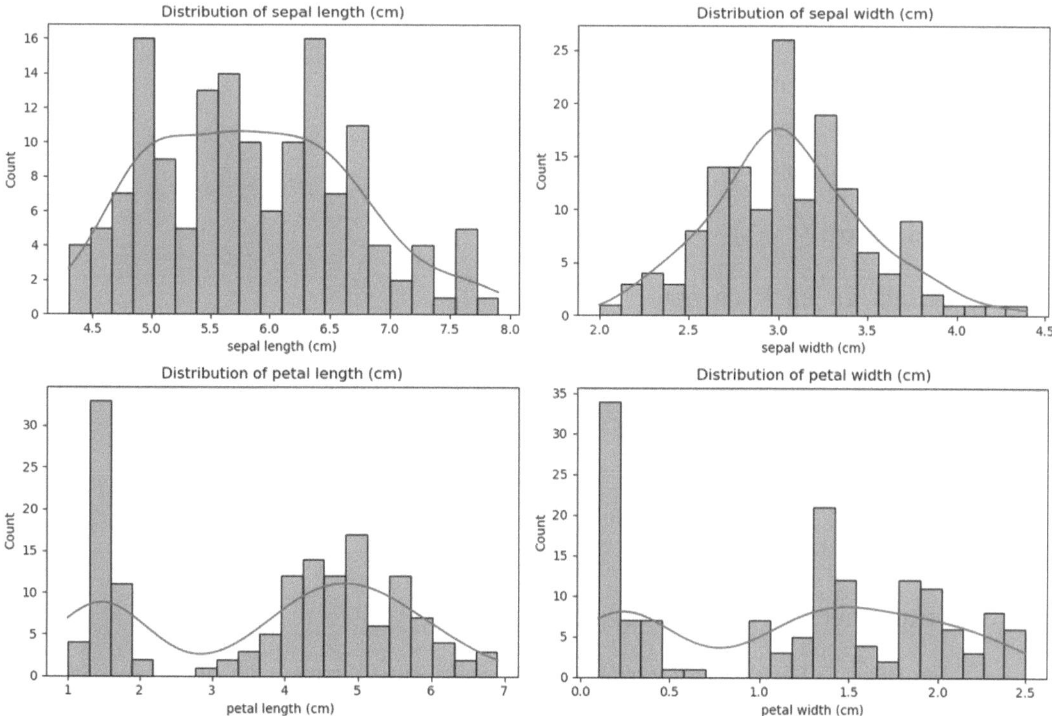

Figure 1-4. *Data Distribution Based on Various Features*

Expected Result

Step 4: Data Preprocessing

This step is for the quality and consistency of input data. It includes checking NA values, partitioning the features and label, label encoding, train-test-splitting, and rescaling features. These steps standardize the data to a point for more comfortable training. The first code snippet shows how to check for missing values, and the second one shows how to handle features.

The code provided checks all the columns in the `DataFrame` for investigating the missing values using `df.isnull().sum()`, ensuring that there are no significant missing values before proceeding with model training.

Listing 1-13. Checking Missing Data

```
# Checking for missing values
print("Missing values in each column:")
print(df.isnull().sum())
```

Expected Result

```
Missing values in each column:
sepal length (cm)    0
sepal width (cm)     0
petal length (cm)    0
petal width (cm)     0
species              0
dtype: int64
```

We have another step here. In this step:

- Features (X) and target variable (y) are separated from each other.
- `LabelEncoder` converts the categorical species names into numeric values.
- The dataset is split into training and testing sets by an 80-20 ratio.
- The `StandardScaler` standardizes the features so that each feature will have an equal contribution to model training.

Also, this step does not have any output.

Listing 1-14. Step 4: Data Preprocessing

```
# Separating features and target variable
X = df.drop('species', axis=1)
y = df['species']

# Encoding target labels
from sklearn.preprocessing import LabelEncoder
le = LabelEncoder()
y_encoded = le.fit_transform(y)
```

```
# Splitting the dataset into training and testing sets
X_train, X_test, y_train, y_test = train_test_split(X, y_encoded, test_
size=0.2, random_state=42)

# Feature scaling
scaler = StandardScaler()
X_train_scaled = scaler.fit_transform(X_train)
X_test_scaled = scaler.transform(X_test)
```

Step 5: Model Training

In this step, you will train a Random Forest classifier. This procedure ensembles a learning method and is highly required for solving classification problems. It improves performance and reduces overfitting by averaging the predictions of multiple decision trees. The goal of this step is to build an accurate model that can precisely classify iris species based on the provided input features. The code below trains the model. This code creates a `RandomForestClassifier` with 100 estimators and fits it to the training data. Using a random state guarantees that results can be replicated. When training goes smoothly, a verification message appears.

Listing 1-15. Step 5: Model Training

```
# Initializing the Random Forest Classifier
rf_classifier = RandomForestClassifier(n_estimators=100, random_state=42)

# Training the model
rf_classifier.fit(X_train_scaled, y_train)

print("Model training completed.")
```

Expected Result

```
Model training completed.
```

Step 6: Model Evaluation

Understanding the performance of your classifier depends on model evaluation. Using the trained model, you can make predictions on the test set. This step assesses the performance using metrics such as accuracy, classification report, and confusion matrix.

This thorough assessment aids in determining the model's generalization of unseen data. The first code snippet below calculates the accuracy, and the second one shows how to generate a detailed classification report and visualize the confusion matrix. In the below code, the `predict()` method generates predictions on the scaled test data. The `accuracy_score` function calculates the ratio of correct predictions, providing a straightforward metric of model performance.

Listing 1-16. Step 6: Model Evaluation

```
# Making predictions on the test set
y_pred = rf_classifier.predict(X_test_scaled)

# Calculating accuracy
accuracy = accuracy_score(y_test, y_pred)
print(f"Accuracy of the model: {accuracy * 100:.2f}%")
```

Expected Result

```
Accuracy of the model: 100.00%
```

The below code provides a classification report. The classification report provides additional performance metrics such as precision, recall, F1 score, and support for each class.

Listing 1-17. Step 6: Model Evaluation

```
# Detailed classification report
print("Classification Report:")
print(classification_report(y_test, y_pred, target_names=iris.
target_names))
```

Expected Result

```
Classification Report:
              precision    recall  f1-score   support

      setosa       1.00      1.00      1.00        10
  versicolor       1.00      1.00      1.00         9
   virginica       1.00      1.00      1.00        11
```

```
    accuracy                           1.00        30
   macro avg       1.00      1.00      1.00        30
weighted avg       1.00      1.00      1.00        30
```

`confusion_matrix`: Summarizes prediction results by showing the counts of true vs. predicted labels. Figure 1-5 shows the confusion matrix.

Listing 1-18. Confusion Matrix Visualization

```
conf_matrix = confusion_matrix(y_test, y_pred)
plt.figure(figsize=(6,4))
sns.heatmap(conf_matrix, annot=True, fmt='d', cmap='Blues',
            xticklabels=iris.target_names,
            yticklabels=iris.target_names)
plt.ylabel('Actual')
plt.xlabel('Predicted')
plt.title('Confusion Matrix')
plt.show()
```

Expected Result

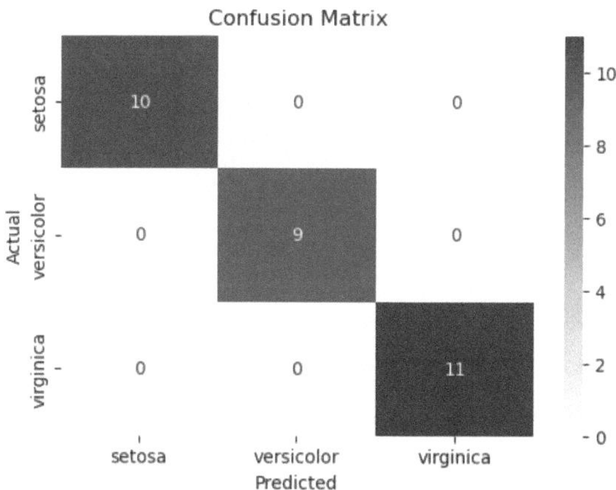

Figure 1-5. *Confusion Matrix*

Step 7: Model Deployment (Simulated)

In a real-world scenario, deploying an ML model is a way to make the model accessible for predictions. In this project, deployment is simulated by creating a prediction function. We make a function that accepts input features, applies the same scaling as the training data, and then returns the predicted output. This step illustrates how the trained model can be encapsulated into a reusable component. In real-world projects, the model and function can run on a remote server, and users can access them using the API. The code below creates a `dataFrame` for input features.

The `predict_species` function accepts four measurements of an iris flower as inputs. It creates a `DataFrame` from these values, scales them, and then passes the data into the trained model for making predictions. Finally, the code converts the numeric prediction to a human-readable species name. This function demonstrates how you might integrate the model into a real production environment.

Listing 1-19. Step 7: Model Deployment

```
def predict_species(sepal_length, sepal_width, petal_length, petal_width):
        # Creating a DataFrame for the input
        input_data = pd.DataFrame({
        'sepal length (cm)': [sepal_length],
        'sepal width (cm)': [sepal_width],
        'petal length (cm)': [petal_length],
        'petal width (cm)': [petal_width]
    })

    # Scaling the input data
    input_scaled = scaler.transform(input_data)

    # Making prediction`
    prediction = rf_classifier.predict(input_scaled)

    # Mapping numerical prediction to species name
    species = le.inverse_transform(prediction)[0]
    return species
```

```
# Example usage of the prediction function
example_species = predict_species(5.1, 3.5, 1.4, 0.2)
print(f"The predicted species is: {example_species}")
```

Expected Results

```
The predicted species is: setosa
```

Step 8: Visualizing Feature Importance

Knowing which features are most important in the model's decision-making can offer significant prospects. The final step is to extract the feature importance from the Random Forest model and present it in a bar chart. The model will be better understood using this visual representation, which also verifies whether its decisions follow domain knowledge. The code below visualizes feature importance.

The attribute `feature_importances_` of the Random Forest model gives a rating for every feature, showing its significance. The code first arranges these ratings into a `DataFrame`, sorts them, and then uses a bar chart to demonstrate the effect of each feature.

Listing 1-20. Step 8: Visualizing Feature Importance

```
feature_importances = rf_classifier.feature_importances_
features = X.columns

# Creating a DataFrame for visualization
importance_df = pd.DataFrame({
    'Feature': features,
    'Importance': feature_importances
}).sort_values(by='Importance', ascending=False)

# Plotting feature importances
plt.figure(figsize=(8,6))
sns.barplot(x='Importance', y='Feature', data=importance_df,
palette='viridis')
plt.title('Feature Importances in Random Forest Classifier')
plt.xlabel('Importance')
plt.ylabel('Feature')
plt.show()
```

Expected Results

Figure 1-6 indicates which features (such as petal length or sepal width) have the highest effect on the model's predictions. This step is important in explaining how the final model predicts. It is a part of explainability that we will talk about in Chapter 9. Here, I just wanted to make you familiar with it.

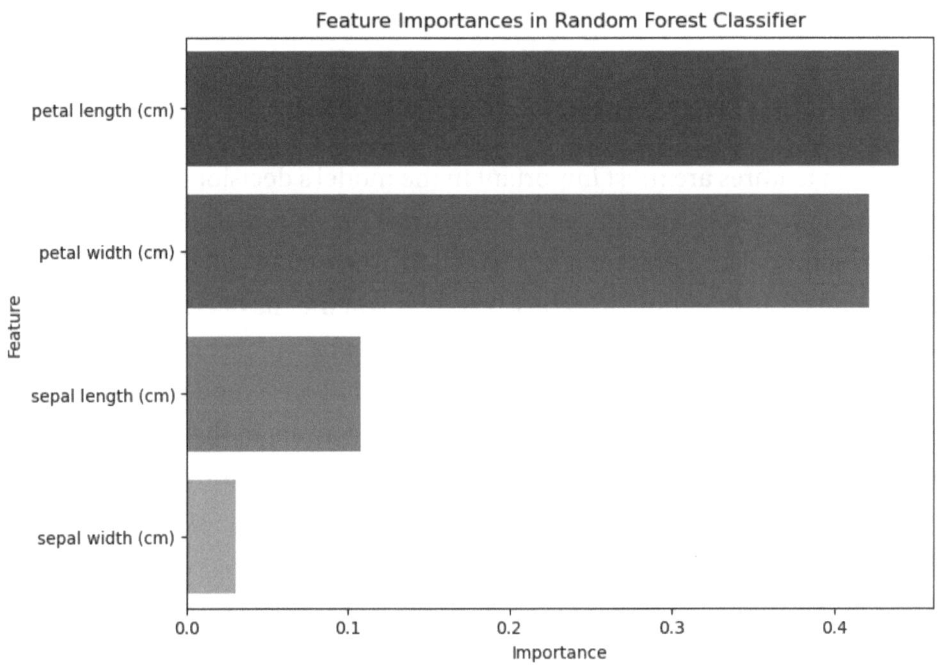

***Figure 1-6.** Bar Chart for Showing the Effect of Each Feature on Model Prediction*

Conclusion

In this project, we successfully created a general-purpose machine learning pipeline that will be your foundation for the entire lifecycle of an ML model with basic components. This is a good starting point for tougher topics in future chapters.

Summary

This chapter focused on the basics of machine learning architecture, including definitions, components, and relationships. ML architecture is just the backbone of a successful machine learning system, offering a clean and scalable approach to create, train, deploy, and operate on models. We discussed challenges (such as scaling, latency, versioning, operational complexity, and data bias) and showed how cloud solutions, MLOps, and automation help to overcome these challenges. We also had an extensive talk about the iterative machine learning lifecycle. The importance of grasping the stages of the lifecycle—from problem definition to deployment and monitoring—is critical for delivering stable ML systems that can keep their performance up with changing data and business landscape. This chapter introduced the foundational concepts of ML architecture. In the upcoming chapters, we will dive deeper into these areas. The first step in developing effective machine learning models is preparing a high-quality dataset. The next chapter provides an in-depth discussion of data pipelines.

CHAPTER 2

Data Pipeline Design for Machine Learning

The heart of any successful machine learning (ML) project is an efficient and powerful data pipeline. A data pipeline guarantees that data flows seamlessly through all the phases of the ML lifecycle. These phases include data collection and preprocessing, training models, as well as deployment. In this chapter, we cover the conceptual pieces and guidelines for creating data pipelines that are ML-centric. We start with data collection and cleaning. We will have a detailed discussion about various techniques of data cleaning, and we will cover the best practices. Next, we provide an in-depth exploration of feature engineering, starting from fundamental concepts and progressing to more advanced topics. We also address working with text data, which often requires additional preprocessing steps.

Another real-world problem is imbalanced data. This frequently happens in practical problems. For instance, in medical science, the number of cases in which a disease is diagnosed is considerably less than the number of healthy people. Dealing with these types of data needs its own strategy, which will be covered in this chapter. Finally, we talk about scaling and distributed systems in data analysis. As machine learning projects are getting bigger, using distributed systems to handle such big data is inevitable. As with previous chapters, we conclude with a hands-on project to reinforce and apply the concepts covered here.

Data Collection and Cleaning

The first step for developing machine learning models, assuming you have an insight into which problem you are going to solve, is data collection and cleaning. This step provides the data, cleans it, and prepares it for the next step. Here, first we will discuss data collection, and then we will proceed with data cleaning.

M. R. Mahdiani, *Mastering Machine Learning Architecture and Solutions*,
https://doi.org/10.1007/979-8-8688-2527-9_2

Data Collection

In an ML pipeline, data collection is the first step. The goal of data collection is to collect data from multiple sources, prepare it, and feed it into a predictive model. Data can be collected from various sources, like

- **Internal Databases**: Transactional databases, CRM systems, and business intelligence tools
- **Open Datasets**: Government archives, open-data platforms (e.g., Kaggle, UCI Machine Learning Archive, et al.)
- **Web Scraping**: Scrape the data from publicly available web pages using tools like Beautiful Soup or Scrapy
- **APIs**: APIs for fetching real-time data from Twitter, financial data providers, or weather services

Below is a sample code that shows how to pull data from an API.

Listing 2-1. Sample API Endpoint

```
import requests
import pandas as pd

# Define the API endpoint and parameters
api_url = "https://api.openweathermap.org/data/2.5/weather"
params = {
    'q': 'London',
    'appid': 'YOUR_API_KEY'
}

# Make the API request
response = requests.get(api_url, params=params)

# Check if the request was successful
if response.status_code == 200:
    data = response.json()
    # Convert the relevant data to a pandas DataFrame
    weather_data = pd.DataFrame([data])
    print(weather_data)
```

```
else:
    print(f"Failed to retrieve data: {response.status_code}")
```

Expected Results

If you do not enter a correct API key, you will see this:

```
Failed to retrieve data: 401
```

However, by having a valid API key, you can access and download the data. Figure 2-1 illustrates various types of data collection. In this figure, multiple data sources are fed into a central data repository for subsequent processing.

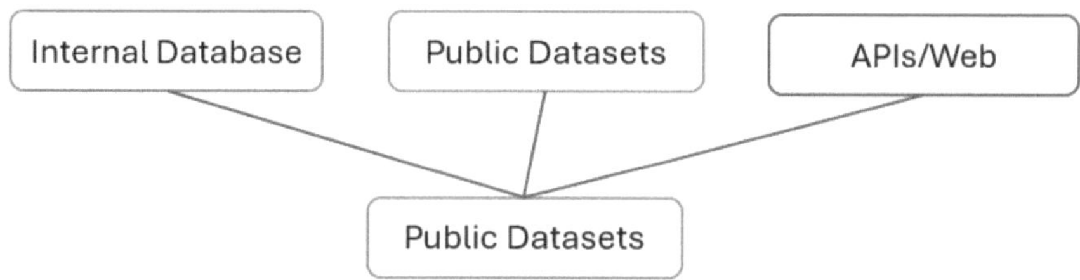

Figure 2-1. *Data Collection Process*

Data Cleaning

After data collection, data must be cleaned for inconsistencies, null/blank values, and inaccurate data. Good data quality is key for enhancing both the accuracy and reliability of the ML model. In the project of Chapter 1, we had a brief talk about data cleaning; here, we will discuss data cleaning in more detail.

Handling Missing Values

In real-world datasets, missing data is common. There are a few methodologies to deal with missing values:

- **Removal**: If the number of missing information is slim, then the rows or columns containing such data can simply be removed and will not substantially affect the model.
- **Imputation**: Replace missing values with a statistical estimation such as mean, median, or mode.

Below is an example of imputing missing data in a column with the median.

Listing 2-2. Impute Missing Values Using the Median

```
# Impute missing values using the median
from sklearn.impute import SimpleImputer
import pandas as pd

# Sample dataset
data = {'age': [25, 30, None, 35, 40], 'salary': [50000, 60000, 55000,
None, 70000]}
df = pd.DataFrame(data)

# Create an imputer object
imputer = SimpleImputer(strategy='median')

# Apply imputation
clean_data = pd.DataFrame(imputer.fit_transform(df), columns=df.columns)
print(clean_data)
```

Expected Results

```
   age   salary
1  25.0  50000.0
1  30.0  60000.0
2  32.5  55000.0
3  35.0  57500.0
4  40.0  70000.0
```

Removing Duplicates

Duplicate entries can bias the model and lead to misleading insights. Below is the Pandas method for removing duplicates:

```
# Remove duplicate rows
df = df.drop_duplicates()
```

Handling Outliers

Outliers can negatively affect a model and lead to poor performance. Some common approaches to deal with outliers are

- **Z-Score Method**: Dropping data points that are outside a certain number of standard deviations from the mean.
- **IQR Method**: Use the interquartile range to detect and filter extreme values outside the typical spread of the data.

A summary of the most common cleaning techniques is represented in Table 2-1.

Table 2-1. *A Summary of Some Commonly Used Data Cleaning Techniques and When to Use Them*

Technique	Description	Use Case
Removing Duplicates	Remove repeated data entries	When there are unintentional copies
Imputation	Fill missing values with median, mean, etc.	When missing data is minimal
Handling Outliers	Remove or cap outliers	When outliers skew the model

Best Practices for Data Collection and Cleaning

Data collection and cleaning are essential steps in a data-based project. This helps us to gain accurate analysis, build good models, and derive in-depth insights into the data. In the following, we discuss the best practices for data collection and cleaning.

Automate Data Validation

Manual validation is prone to mistakes and is time-consuming. To ensure consistency and scalability of the process, use scripts (e.g., Python with libraries such as Pandas or Great Expectations to validate your data against defined rules like data type, range, etc.). Overall, try to automate any step if it is possible. Automation gives a higher degree of confidence and also frees your time for more important tasks. The only exception is when a project is so small that setting up an automated pipeline would take longer than performing the task manually. Below is a sample code for automating data validation.

Listing 2-3. Automate Data Validation

```
import pandas as pd

def validate_data(df):
    # Check for missing values
    assert df.isnull().sum().sum() == 0, "Missing values detected!"

    # Validate data types
    assert df['age'].dtype == int, "Age column must be of type integer."

    # Ensure values fall within expected ranges
    assert df['age'].between(0, 120).all(), "Age values out of range."
```

Document Data Sources

Documentation is paramount to keep the process transparent and reproducible and provides an ease of collaboration for the team members. Keep a data dictionary for specifying the data of each field, its goal, and its format. Also, document the data source (i.e., APIs, databases, or third party) and frequencies at which materials are updated in the docs. For example:

- **Data Source**: Customer transaction data from Shopify API
- **Frequency**: Updated hourly
- **Fields**:
    - `order_id`: Unique identifier for each order (type: string)
    - `order_date`: Date of the order (type: datetime)
    - `total_amount`: Total order amount in USD (type: float)

Documenting the data sources is the first step for data governance and reproducibility. More scientifically, we store the metadata of each dataset in Python dictionaries to keep it simple. The following code shows how you can store data information (metadata).

Listing 2-4. Documenting Data Sources

```
data_sources = {
    "customer_data": {
```

```
        "name": "Customer Database",
        "location": "/path/to/customer_data.csv",
        "format": "CSV",
        "update_frequency": "Daily",
        "description": "Contains customer demographics and purchase
        history."
    },
    "product_catalog": {
        "name": "Product Catalog API",
        "endpoint": "https://api.example.com/products",
        "format": "JSON",
        "update_frequency": "Real-time",
        "description": "Provides information about available products."
    },
    "sales_data": {
        "name": "Sales Transactions",
        "location": "s3://my-bucket/sales_data/",
        "format": "Parquet",
        "update_frequency": "Hourly",
        "description": "Records of sales transactions."
    }
}

# Accessing information about a specific data source
print(data_sources["customer_data"]["name"])
print(data_sources["product_catalog"]["endpoint"])

# Iterating through all data sources
for source_name, metadata in data_sources.items():
    print(f"\nSource: {source_name}")
    for key, value in metadata.items():
        print(f"  {key}: {value}")
```

Expected Results

If you run the above code, you will see the following results.

```
Customer Database
https://api.example.com/products

Source: customer_data
  name: Customer Database
  location: /path/to/customer_data.csv
  format: CSV
  update_frequency: Daily
  description: Contains customer demographics and purchase history.

Source: product_catalog
  name: Product Catalog API
  endpoint: https://api.example.com/products
  format: JSON
  update_frequency: Real-time
  description: Provides information about available products.

Source: sales_data
  name: Sales Transactions
  location: s3://my-bucket/sales_data/
  format: Parquet
  update_frequency: Hourly
  description: Records of sales transactions.
```

Monitor Data Quality

Data quality can degrade over time due to changes in data sources, updates in collection methods, or other external issues. Continuous monitoring makes it possible to identify the issues and fix them in the early stages. Consider two approaches:

- Create automated alerts for anything unusual (say, a precipitous drop in the volume of data or some unexpected nulls).
- Employ tools such as DataDog or Monte Carlo to monitor KPIs (key performance indicators), such as precision, quality, and timeliness.

Here is a code example to simulate the logic of the data quality check.

Listing 2-5. Data Quality Check

```
import random

def check_missing_values(data):
    missing_count = sum(1 for x in data if x is None)
    return missing_count

def check_data_range(data, min_val, max_val):
    out_of_range_count = 0
    for x in data:
        if x is not None and isinstance(x, (int, float)) and not (min_val
        <= x <= max_val):  # Check type *before* comparing
            out_of_range_count += 1
    return out_of_range_count

def check_data_types(data, expected_type):
    incorrect_type_count = 0
    for x in data:
        if x is not None and not isinstance(x, expected_type):
            incorrect_type_count += 1
    return incorrect_type_count

# Sample data
temperature_data = [25, 22, None, 28, 31, 150, 27, None, 29, "abc",
30.5]  # Includes missing, out-of-range, incorrect type, and float

# Performing data quality checks
missing_values = check_missing_values(temperature_data)
out_of_range_values = check_data_range(temperature_data, 0, 100)
incorrect_types = check_data_types(temperature_data, int)

print(f"Missing Values: {missing_values}")
print(f"Out of Range Values: {out_of_range_values}")
print(f"Incorrect Type Values: {incorrect_types}")
```

Expected Results

```
Missing Values: 2
Out of Range Values: 1
Incorrect Type Values: 2
```

We will dive deep into efficient monitoring techniques in Chapter 6.

Standardized Data Formats

Taking inconsistent formats (dates, currencies, or units) will make later analysis confusing and may even lead to incorrect results. Establish and impose data standards (ISO 8601 for dates, USD currency) as well as transform data into one consistent form while cleaning using transformation scripts.

Data standardization, especially dates (among other data formats), is extremely important for data consistency. In the code below, we use the Pandas library to normalize date strings.

Listing 2-6. Standardize Data Formats

```
import pandas as pd

# Sample data with different date formats
date_strings = ["2023-10-26", "10/26/2023", "Oct 26, 2023"]

# Convert to datetime objects with a specified format
dates = pd.to_datetime(date_strings, errors='coerce') #errors='coerce' will
convert invalid parsing into NaT.

# Format the dates consistently (e.g., YYYY-MM-DD)
formatted_dates = dates.strftime('%Y-%m-%d')

print("Original Dates:", date_strings)
print("Formatted Dates:", formatted_dates)
```

Expected Results

```
Original Dates: ['2023-10-26', '10/26/2023', 'Oct 26, 2023']
Formatted Dates: Index(['2023-10-26', nan, nan], dtype='object')
```

Dates are very tricky. Ensure that every date format used in your dataset is correctly interpreted by your code. When formats are uncertain or mixed, it is a good practice to write a helper function to safely convert all values to datetime. I consistently use this practice, and it is very helpful in big projects.

Leverage Version Control for Data

The evolution of a dataset over time is necessary to track, and this ensures reproducibility and supports debugging. Tools such as DVC (Data Version Control) or Delta Lake are used to store a changelog of the data, updates, fixes, and enhancements.

Collaboration with Domain Experts

Domain experts provide essential context that helps interpret, validate, and clean the data. Collaborate with them to define rules for data validation and identify important features. Regularly reviewing data quality reports with experts ensures that the data aligns with real-world expectations. The code below simulates the effect of talking with an expert to improve data quality.

Listing 2-7. Collaborating with Domain Experts

```
medical_data = [
    {"patient_id": 1, "symptoms": ["fever", "cough"], "diagnosis": None,
    "expert_feedback": None},
    {"patient_id": 2, "symptoms": ["headache", "fatigue"], "diagnosis":
    "Migraine", "expert_feedback": None},
    {"patient_id": 3, "symptoms": ["chest pain", "shortness of breath"],
    "diagnosis": "Possible Heart Condition", "expert_feedback": None},
]

# Simulate a discussion with a domain expert
def collaborate_with_expert(data):
    for record in data:

        # Case 1: No diagnosis yet
        if record["diagnosis"] is None:
            if "fever" in record["symptoms"] and "cough" in
            record["symptoms"]:
```

```
                record["diagnosis"] = "Flu"
                record["expert_feedback"] = "Consistent with seasonal flu."
            Else:
                record["diagnosis"] = "Unknown"
                record["expert_feedback"] = "Further investigation needed."

        # Case 2: Diagnosis already exists
        else:
            if record["diagnosis"] == "Possible Heart Condition" and "chest
            pain" in record["symptoms"]:
                record["expert_feedback"] = "Likely heart condition,
                further tests required."
            Else:
                record["expert_feedback"] = "Diagnosis confirmed."
    Return data

updated_data = collaborate_with_expert(medical_data)

for record in updated_data:
    print(record)
```

Expected Results

```
{'patient_id': 1, 'symptoms': ['fever', 'cough'], 'diagnosis': 'Flu',
'expert_feedback': 'Consistent with seasonal flu.'}
{'patient_id': 2, 'symptoms': ['headache', 'fatigue'], 'diagnosis':
'Migraine', 'expert_feedback': 'Diagnosis Confirmed'}
{'patient_id': 3, 'symptoms': ['chest pain', 'shortness of breath'],
'diagnosis': 'Possible Heart Condition', 'expert_feedback': 'Likely heart
condition, further tests required'}
```

Many codes in this book are simulations. So, please read the code line by line to get a better insight into the problems and their solutions. In real-world scenarios, discussions with experts typically result in clear and formalized validation rules.

Test and Iterate

Data cleaning is a step-by-step task. Testing is essential to ensure that cleaning operations are correct and do not introduce new issues. Use small validation checks after each cleaning step, and review the data again to catch any remaining problems. In practice, you repeatedly check and clean the data until no further issues are detected. The code below shows some automated tests that you can run each time after cleaning your data.

Listing 2-8. Test and Iterate (Data Quality)

```
import unittest  # Using the unittest framework for testing

# (Data quality check functions from previous examples would be here)

class TestDataQuality(unittest.TestCase):
    def test_missing_values(self):
        data = [1, 2, None, 4, None]
        self.assertEqual(check_missing_values(data), 2)

    def test_data_range(self):
        data = [10, 50, 150, 20, None]
        self.assertEqual(check_data_range(data, 0, 100), 1)

    def test_data_types(self):
        data = [1, 2, "abc", 4, None]
        self.assertEqual(check_data_types(data, int), 1)

    # Add more test cases as needed...

if __name__ == '__main__':
    unittest.main(argv=['first-arg-is-ignored'], exit=False) # To run the
    tests within the interactive environment
```

Expected Results

```
...
Ran 3 tests in 0.002s

OK
```

When data is gathered and cleaned, we can proceed to the next step, which is feature engineering. Feature engineering generates meaningful input features that will be used by the machine learning model.

Feature Engineering Techniques

Feature engineering is the key to the machine learning process. It strongly influences model performance. By changing raw data into clear features, you help machine learning techniques analyze the data better.[5,6] In this section, we'll talk about various ways to extract, create, and transform features that make the data input better.

Feature Extraction

Feature extraction works by turning raw facts into features that more clearly show the main issue. This helps with model learning. Extracting features, especially in text and images, is difficult.[6–8] For these cases, the usual ways to extract features include

- **Text Feature Extraction**: Converting text into a numerical form using methods such as TF-IDF or word embeddings
- **Image Feature Extraction**: Using convolutional neural networks (CNNs) to extract useful patterns from images

Later in this chapter, we will talk more about handling features in text files.

Feature Creation

Feature creation is about making new features that may improve model performance. One of the most common techniques is polynomial feature generation, where new features are created as polynomial combinations of the original ones.

For instance, making polynomial features by using the `PolynomialFeatures` class from `sklearn`. The code below dives into it.

Listing 2-9. Polynomial Feature

```
from sklearn.preprocessing import PolynomialFeatures
import numpy as np

# Sample data
X = np.array([[2, 3], [3, 4], [5, 6]])

# Create polynomial features (degree 2)
poly = PolynomialFeatures(degree=2)
X_poly = poly.fit_transform(X)
print(X_poly)
```

Expected Results

```
[[ 1.  2.  3.  4.  6.  9.]
 [ 1.  3.  4.  9. 12. 16.]
 [ 1.  5.  6. 25. 30. 36.]]
```

Another technique is to extract additional information from datetime columns such as the day of the week, month, or hour to capture seasonal trends. The code below shows an example of extracting date features.

Listing 2-10. Extracting Date Features

```
import pandas as pd
import matplotlib.pyplot as plt

# 1. Create sample data with datetime and sales
df = pd.DataFrame({
    'datetime': pd.date_range('2023-01-01', '2023-12-31', freq='D'),
    'sales': [100 + (i % 30) * 5 for i in range(365)]  # Synthetic
    seasonal pattern
})

# 2. Extract datetime features
df['day_of_week'] = df['datetime'].dt.dayofweek  # 0=Monday
df['month'] = df['datetime'].dt.month
df['hour'] = df['datetime'].dt.hour  # (Will be 0 since daily data)
df['week_of_year'] = df['datetime'].dt.isocalendar().week
```

```
# 3. Analyze trends
print("Average sales by day of week:")
print(df.groupby('day_of_week')['sales'].mean())

print("\nAverage sales by month:")
print(df.groupby('month')['sales'].mean())

# 4. Visualize monthly trend
df.groupby('month')['sales'].mean().plot(
    kind='bar',
    title='Seasonal Trend: Average Sales by Month',
    xlabel='Month',
    ylabel='Sales'
)
plt.show()
```

Expected Results

```
Average sales by day of week:
day_of_week
1     172.500000
1     171.730769
2     170.961538
3     170.192308
4     172.307692
5     171.538462
6     172.264151
Name: sales, dtype: float64

Average sales by month:
month
1      170.161290
2      172.500000
3      174.838710
4      172.500000
5      170.161290
6      172.500000
7      170.322581
```

```
8       170.483871
9       172.500000
10      170.645161
11      172.500000
12      170.806452
Name: sales, dtype: float64
```

In addition to the numeric output, above code generates the chart of Figure 2-2 which illustrates the seasonality.

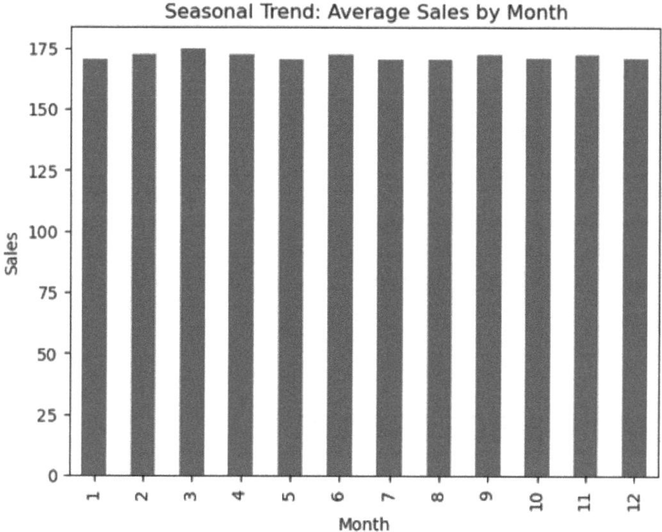

***Figure 2-2.** Average Sales and Seasonality*

Talking with experts can help you decide whether seasonal patterns should be included as features.

Feature Transformation

Feature transformation techniques improve the model's accuracy. The most widely used ones include the following:

> **Normalization and Standardization**: Rescaling the data so that the features have the same midpoints in the corresponding axes is beneficial since algorithms will converge more quickly. For example, `StandardScaler` from `scikit-learn` standardizes features by removing the mean and scaling to unit variance. The code below shows how you could transform the features.

Listing 2-11. Feature Transformation

```
from sklearn.preprocessing import StandardScaler

# Sample data
X = [[100, 0.1], [200, 0.2], [300, 0.3]]

# Standardize features
scaler = StandardScaler()
X_scaled = scaler.fit_transform(X)
print(X_scaled)
```

Expected Results

```
[[-1.22474487e+00 -1.22474487e+00]
 [ 0.00000000e+00 -3.39934989e-16]
 [ 1.22474487e+00  1.22474487e+00]]
```

Another approach is the log transform, which is used for decreasing the skewness of data distributions.

Handling Complicated Features

There are many cases of machine learning scenarios where one or more parameters are text. Working with texts is not as easy as other types of data. Here in this section, we will have a quick discussion about handling text in machine learning. Some examples of dealing with text include natural language processing (NLP), search engines, and sentiment analysis. This section covers necessary techniques for text preprocessing, including tokenization, stemming, and lemmatization. In addition, advanced methods for text representation, like word embeddings and TF-IDF, will be discussed here.

Text Preprocessing

Text data is inherently unstructured and noisy. We cannot use text directly in our models. We need to preprocess the text extensively before being able to use it in machine learning models. Preprocessing converts raw text into an analyzable and clean format, which we need for our deep learning techniques or our traditional machine learning models.

Tokenization

The first step for preprocessing the text is to tokenize it. Tokenization means breaking a text into smaller units. Each unit is called a token. These units can be sentences, words, or sub-words. Here, we check different types of tokenization.

First, we check word tokenization. The following code shows how to split text into individual words.

Listing 2-12. Splitting Text into Individual Words

```
# Before Running the code for the first time you need to download NLTK
tokenizer

import nltk
nltk.download('punkt_tab')

# Then proceed with tokenizer

from nltk.tokenize import word_tokenize

text = "Machine learning is fascinating."
Tokens = word_tokenize(text)
print(tokens)
```

Expected Results

```
# ['Machine', 'learning', 'is', 'fascinating', '.']
```

If you decide to tokenize the text using sentences, you can use the following code.

Listing 2-13. Splitting Text into Sentences

```
from nltk.tokenize import sent_tokenize
text = "Machine learning is fascinating. It has many applications."
Sentences = sent_tokenize(text)
print(sentences)
```

Expected Results

```
# ['Machine learning is fascinating.', 'It has many applications.']
```

In addition to words and sentences, we mentioned that tokens can be sub-words. Sub-words are a good selection for handling rare words (e.g., Byte Pair Encoding in transformers).

In addition to tokenization, there are some other techniques that we should consider for text analysis. After tokenization, we should consider stemming. Stemming gets the roots of the words by removing suffixes. It is a rule-based heuristic approach. In the code below, you can see an example of stemming.

Listing 2-14. Sample Code for Stemming

```
from nltk.stem import PorterStemmer

stemmer = PorterStemmer()
words = ["running", "runner", "runs"]
stems = [stemmer.stem(word) for word in words]
print(stems)
```

Expected Results

```
# ['run', 'runner', 'run']
```

The advantage of stemming is that it reduces computation complexity and vocabulary size. However, it can produce non-dictionary words (e.g., "running" to "run"). So, rule-based alone is not enough for getting the root of a word. Here, lemmatization can help. Lemmatization uses a dictionary form (lemma) to map words to their roots. It is more accurate than stemming but computationally expensive. The code below shows an example of lemmatization.

Listing 2-15. Sample Code for Lemmatization

```
# If you have not downloaded yet, download wordnet first
import nltk
nltk.download('wordnet')

# Then proceed with the following
from nltk.stem import WordNetLemmatizer
lemmatizer = WordNetLemmatizer()
words = ["running", "runner", "runs"]
lemmas = [lemmatizer.lemmatize(word, pos="v") for word in words]
print(lemmas)
```

Expected Results

```
# ['run', 'runner', 'run']
```

In the above code, you can see that lemmatize needs the part of speech (here, verb) as an input, which makes it a little more complicated to use. Figure 2-3 compares the processes of stemming and lemmatization. This figure illustrates their differences in methodology and outputs. Stemming produces root forms, and lemmatization yields dictionary-valid lemmas.

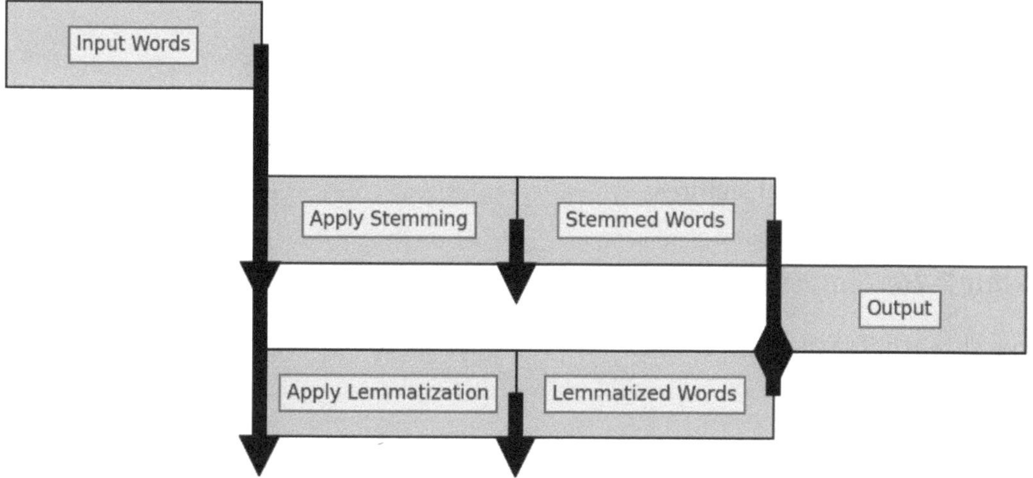

Figure 2-3. *Flowchart Comparing the Processes of Stemming and Lemmatization*

Text Representation

After the above steps, we need to convert the text to numerical formats to be able to use that in our machine learning models. There are some techniques for this conversion. One of the most popular ones is TF-IDF. In the next section, we will talk about TF-IDF.

TF-IDF (Term Frequency-Inverse Document Frequency)

TF-IDF considers the significance of a word in a document relative to the collection of words, which is called a corpus.

$$TF-IDF(t,d) = TF(t,d) \times IDF(t)$$

In the above formula:

- **TF**: Term frequency of term t in document dd
- **IDF**: Inverse document frequency, calculated as

$$IDF(t) = log\frac{N}{n_t},$$

where

N = is the total number of documents in the corpus

n_t = is the number of documents containing term t

- The code below shows a technique to use TF-IDF to change text facts into numerical features.

Listing 2-16. Simple TF-IDF

```
from sklearn.feature_extraction.text import TfidfVectorizer

# Sample text data
text_data = ["Machine learning is amazing", "Feature engineering is key to
model performance"]

# Create a TF-IDF Vectorizer
vectorizer = TfidfVectorizer()

# Transform the text data
tfidf_features = vectorizer.fit_transform(text_data)
print(tfidf_features.toarray())
```

Expected Results

```
[[0.53404633 0.         0.         0.37997836 0.         0.53404633
  0.53404633 0.         0.         0.        ]
 [0.         0.39204401 0.39204401 0.27894255 0.39204401 0.
  0.         0.39204401 0.39204401 0.39204401]]
```

Word Embeddings

In addition to TF-IDF, there is another technique for converting words to numerical text. One of these methods is word embedding. Word embeddings map words or sequences of words to dense numerical vectors that capture semantic meaning. There are several techniques for word embedding, such as Word2Vec, GloVe, and FastText. Word2Vec uses a neural network to learn word representations based on context. Another method is GloVe, which combines global statistical information with local context. And FastText considers sub-word information to better handle rare or out-of-vocabulary words. You can also use pre-trained embeddings for better generalization and faster development (e.g., GloVe, Google Word2Vec). Pre-trained embeddings are very effective, and I have used them in many projects. The main advantage of this technique is that it can understand the semantic relationship between the words. Below shows a sample code for pre-trained embedding.

Listing 2-17. A Sample Code for Using Pre-trained Embedding

```
import enism.downloader as api

model = api.load("glove-wiki-gigaword-50")
print(model["machine"])  # Vector representation of 'machine'
```

Expected Result

```
[-0.34165   -0.81267    1.4513     0.05914   -0.080801   0.39567
  0.10064   -0.5468    -0.18887    0.11364   -0.040956  -0.5637
 -0.32191    0.15968   -0.59756   -0.14571   -0.77074    1.2955
 -0.72002   -0.90818    0.76644    0.05346   -0.0031632 -0.15341
  0.22065   -1.191     -1.0775    -0.29768    1.327     -0.51359
  2.6229    -0.67411   -0.82558    0.14283   -0.014214   0.90775
  0.66828    0.48431    0.1543     0.26044    1.0191     0.015872
 -0.75325    0.58992    0.4546    -0.19678    0.42138   -0.43168
  0.11985    0.14094  ]
```

Contextualized Embeddings

The most advanced technique for text representation is using modern transformers such as BERT, GPT, and RoBERTa. These transformers produce contextualized embeddings, and they evaluate each word with its surrounding words. The code below shows a sample code for using BERT.

Listing 2-18. Sample Code for Using BERT

```
from transformers import AutoTokenizer, AutoModel
import torch

tokenizer = AutoTokenizer.from_pretrained("bert-base-uncased")
model = AutoModel.from_pretrained("bert-base-uncased")

text = "Machine learning is amazing."
Inputs = tokenizer(text, return_tensors="pt")
outputs = model(**inputs)
print(outputs.last_hidden_state.shape)  # Contextualized embeddings
```

Expected Result

```
torch.Size([1, 7, 768])
```

Table 2-2 compares these techniques.

Table 2-2. *Comparison of Text Representation Techniques*

Technique	Key Features	Advantages	Limitations	Examples	Common Applications
TF-IDF	Weighted word importance	Simple, interpretable	Ignores word order and context	N/A	Information retrieval, text mining
Word2Vec	Contextual word relationships	Captures semantic similarity	Requires a large corpus to train	skip-gram, CBOW	Semantic analysis, document clustering
GloVe	Combines local and global context	Pre-trained embeddings available	Static embeddings	N/A	Word similarity, NLP preprocessing
BERT	Contextual embeddings based on transformers	Dynamic, state-of-the-art performance	Computationally expensive	N/A	Question answering, text classification

We discussed several techniques for handling text features in machine learning. Text data is inherently complex, as its structure is unstructured and context-dependent. For example, in sentiment analysis, the feature might be a social media comment (text), and the label could be positive, neutral, or negative.

Similarly, feature engineering for other types of unstructured data is also challenging. Examples of unstructured data include images, audio, video, and time-series data.

Best Practices for Feature Engineering

Feature engineering is key to creating accurate machine learning models. It focuses on converting raw data into descriptive features that can extract the underlying patterns and relationships in the data. Good features alone will not enhance the model's performance. Even with good features, poor design can result in overfitting, data leakage, or unpopular results. Below are the best practices for engineers to follow in the feature engineering process. These techniques help developers make their model more powerful and trustworthy.

Understand the Domain

Domain knowledge is essential for creating meaningful and relevant features. Focusing on the context and subtleties of the data and provision of the implications of the problem can be a great way to create features that can answer that question. For a dataset about health, domain knowledge can give you an idea about new features such as "BMI" (Body Mass Index) from height and weight, or "Aging Group" to classify patients into meaningful categories. Sharing knowledge with the field domain experts as well as finding good industry-specific practices through research is a way of obtaining an understanding of common feature engineering procedures and of making sure that your features are connected to the everyday real situation.

Iterate and Test

Similar to data cleaning, feature engineering is an iterative process that goes through an ongoing cycle of experimentation, evaluation, and refinement. Start with a set of basic features and then measure model performance. Play with new features, say, transformations (e.g., log, square root), interaction (e.g., multiplying two dimensions), or aggregation (e.g., mean, sum). Utilize methods and tools such as feature importance scores (e.g., using tree-based models) or correlation analysis to study each feature's impact. Make sure the features are being assessed in a cross-validation process to ensure they work well with the unseen data. For example, when you are dealing with time-series data analysis, you can iterate by creating features such as moving averages, lag variables, or time-based aggregations (e.g., daily, weekly).[9,10]

Avoid Data Leakage

Data leakage happens when the training process is influenced by the validation or test sets, which results in the overestimation of the performance of the model. To clarify what it means, let's have an example. Assume that you are going to make a model to predict whether customers will terminate their contract or not. You have a comprehensive dataset that includes various data of customers, including age, location, education, start date of contract, termination date of contract, and another field that is the label of the data and says whether customers have terminated their contract or not. Something is fishy! The termination date of the contract should not be there; it is related to the future, and in real-world predictions, you won't know a customer's termination date ahead of time. The model, instead of finding a pattern, will learn that when there is a termination date, the contract is terminated! Remember, in many cases, the data leakage is not as obvious as the above example, and discussing with experts can be helpful in data leakage detection.

To avoid this, make sure that the feature engineering steps are only done on the training set, and then they are applied to the validation and test sets. Do not employ future information (e.g., future sales data) to build features for past observations. A technique like Pipeline in Python's scikit-learn can encapsulate both feature engineering and model training steps; hence, it does not lead to leakage. A case in point is that if you are making a feature such as "average purchase amount," you should calculate it with just the training data and use the same average for the test set.

Focus on Interpretability

Even though incorporating complex features can sometimes boost the model's performance, they often reduce interpretability. Aim to create features that are both useful and understandable. Execute elementary transformations (e.g., scaling and binning) instead of difficult ones (e.g., polynomial features) with only a dramatically improved performance scenario. Analyze the domain together with the domain of interest to come up with meaningful features and explain the rationale for each feature transparently. A better solution would be to build a variable like "price-to-income ratio" that is graspable and follows the main economic principles.

Leverage Automated Feature Engineering Tools

Tools of automation can facilitate the process of feature engineering, particularly in the context of large datasets and tight deadlines. The databases, Feature tools, or TPOT, can be fully automated to propose relations between different data points/records. Both automated and domain-specific features can be employed to reach the desired balance. Thus, for example, Feature tools might propose features automatically, like "total purchases in the last 30 days" or "average time between purchases."

Handle Missing Data Thoughtfully

Missing data is a common problem in feature engineering. The way that you deal with missing data can have a considerable effect on model performance. For numerical features, consider an imputation method such as mean, median, and mode to fill the gaps. For categorical features, you can use an imputation, like the mode. You can also add a "missing" category. Do not remove any rows or columns with missing data except when it is necessary. For instance, in the case that the "income" column has some missing values, replace them with the dataset's median income.

Normalize and Scale Features

Normalization and scaling are the practices that ensure the features are in one scaling range, which is especially important for algorithms that are sensitive to feature magnitudes (e.g., k-neighbor or descent gradients).[11] Common scaling methods are like Min-Max scaling, Standardization (z-score normalization), and Robust Scaling. Use scaling after splitting the data into training and test sets to prevent contamination. The code below shows an example of normalization and scaling features.

Listing 2-19. Normalize and Scale Features

```
import numpy as np
from sklearn.preprocessing import StandardScaler

# Sample feature matrix X
X = np.array([
    [1.0, 10.0, 100.0],
    [2.0, 20.0, 200.0],
    [3.0, 30.0, 300.0],
    [4.0, 40.0, 400.0]
])

# Manually define train/test split for the example
X_train = X[:3]    # first 3 rows = training features
X_test = X[3:]     # last row = test features

scaler = StandardScaler()
X_train_scaled = scaler.fit_transform(X_train)
X_test_scaled = scaler.transform(X_test)

print("X_train_scaled:\n", X_train_scaled)
print("X_test_scaled:\n", X_test_scaled)
```

Expected Result

```
X_train_scaled:
 [[-1.22474487 -1.22474487 -1.22474487]
 [ 0.          0.          0.        ]
 [ 1.22474487  1.22474487  1.22474487]]
X_test_scaled:
 [[2.44948974 2.44948974 2.44948974]]
```

Create Interaction Features

Interaction features encode connections between variables that are not clear by themselves. Multiplying, dividing, or adding variables to generate interaction terms, domain knowledge often guides which interactions are useful. One good example can be dividing the total price by the quantity bought to obtain a feature called "price per unit."

Use Feature Selection Techniques

Not every feature is of the same value and importance. Feature selection helps identify the most important predictors, reduce noise, and improve generalization. It also simplifies the approach and enhances model performance. To have an efficient feature selection, apply methods like feature importance scores from tree-based models, L1 regularization (Lasso), or Recursive Feature Elimination (RFE). Using cross-validation, you can find out how feature selection affects model performance. The code below illustrates feature selection techniques more practically.

Listing 2-20. Feature Selection Example

```
import numpy as np
from sklearn.feature_selection import RFE
from sklearn.ensemble import RandomForestClassifier

# ---- Minimal sample data ----
# 100 samples, 20 features
X_train = np.random.rand(100, 20)

# Binary target variable
y_train = np.random.randint(0, 2, size=100)
```

```
# ---- Feature selection ----
model = RandomForestClassifier()
rfe = RFE(model, n_features_to_select=10)
X_train_selected = rfe.fit_transform(X_train, y_train)

print("Selected feature shape:", X_train_selected.shape)
```

Expected Results

```
Selected feature shape: (100, 10)
```

For more details about the best practices in feature engineering, see Table 2-3.

Table 2-3. *Best Practices for Feature Engineering*

Practice	Description	Tools/Techniques
Understand the Domain	Leverage domain knowledge to create meaningful features.	Collaboration with experts, research
Iterate and Test	Continuously evaluate and refine features for better performance.	Cross-validation, feature importance scores
Avoid Data Leakage	Ensure no information from validation/ test sets leaks into training.	Pipelines, careful feature engineering
Focus on Interpretability	Create features that are both effective and easy to understand.	Simple transformations, documentation
Leverage Automated Tools	Use tools to automate feature generation and save time.	Feature tools, TPOT
Handle Missing Data	Impute or flag missing values thoughtfully.	Mean/median imputation, predictive modeling
Normalize and Scale	Ensure features are on a comparable scale.	Min-Max scaling, Standardization
Create Interaction Features	Capture relationships between variables.	Multiplication, division, domain knowledge
Use Feature Selection	Identify and retain the most relevant features.	RFE, Lasso, feature importance scores

Until now, we have collected data, cleaned it, and found and prepared the features to train the model. The next step is model training. The model training is very expensive and needs a lot of computational resources. As this chapter is about data pipelines, we will not proceed with model training; instead, we will discuss a common case that can happen in data pipelines. It is the case that the data is imbalanced. In the next section, we discuss how to handle imbalanced data.

Handling Imbalanced Data

When we deal with imbalanced data, the number of examples in one class greatly eclipses those in other classes. Particularly in classification tasks, handling imbalanced data is a key issue in machine learning model development. This asymmetry frequently causes biased algorithms that score poorly on minority classes. Imbalanced data issues arise in fraud detection, rare illness diagnosis, and customer churn forecasting, where the minority class stands for significant results.

In this section, we will discuss various strategies to handle imbalanced datasets, including data-level approaches (resampling techniques), algorithm-level approaches (cost-sensitive learning), and appropriate evaluation metrics designed for class imbalance.

Challenges of Imbalanced Data

Handling imbalanced datasets presents these difficulties:

Models tend to bias the majority class, so the minority class suffers from poor performance. For instance, a model that forecasts only the majority class could have high accuracy, but its performance on the underrepresented class could be nearly zero.

Misleading Metrics: For imbalanced sets, standard metrics like accuracy can be deceptive. High accuracy could just mean the majority class is forecasted properly, whereas the minority class is not. For clarification, assume that we have imbalanced data of 98 healthy individuals and 2 patients. The purpose is to identify sickness. If we make a model to tag everyone as healthy, the model would tag 98 people correctly, and its accuracy would be 98%; however, clearly, the model cannot predict well.

Data Sparsity: The minority category might be quite small, which would give the model too little information for meaningful patterns to be learned for those events.

Various techniques deal with imbalanced data problems. These techniques can be grouped into two main groups: data-level and algorithm-level. In the next sections, we will discuss each in more detail.

Data-Level Techniques for Handling Imbalanced Data

One of the most common methods of fixing imbalanced data problems is modifying the dataset to give it more balance. For this purpose, these data-level approaches could be applied.

1. Random Undersampling

By randomly removing samples until the dataset is more balanced, random undersampling reduces the size of the majority class. Though this method is simple to use, it may cause data loss, especially if valuable cases are deleted. The code below shows an example of random under sampling.

Listing 2-21. Random Undersampling

```
from imblearn.under_sampling import RandomUnderSampler
from collections import Counter
from sklearn.datasets import make_classification

# Sample Imbalanced Data
X, y = make_classification(n_samples=100, n_features=2, n_informative=2,
                           n_redundant=0, n_repeated=0, n_classes=2,
                           n_clusters_per_class=1,
                           weights=[0.95, 0.05], # Imbalance: 95% class 0,
                           5% class 1
                           class_sep=0.8, random_state=0)

# Apply Random Undersampling
rus = RandomUnderSampler(random_state=42)
X_resampled, y_resampled = rus.fit_resample(X, y)

print("Original class distribution:", Counter(y))
print("Resampled class distribution:", Counter(y_resampled))
```

Expected Results

```
Original class distribution: Counter({0: 95, 1: 5})
Resampled class distribution: Counter({0: 5, 1: 5})
```

2. Random Oversampling

Random oversampling boosts the number of samples in the minority group by duplicating current examples until the class distribution is even. This method can result in overfitting since the same samples are repeated, and therefore the model is less general. Still, it preserves information. The code below shows an example of oversampling.

Listing 2-22. Random Oversampling for Imbalanced Datasets

```
from imblearn.over_sampling import RandomOverSampler
from collections import Counter
from sklearn.datasets import make_classification

# Sample Imbalanced Data
X, y = make_classification(n_samples=100, n_features=2, n_informative=2,
                           n_redundant=0, n_repeated=0, n_classes=2,
                           n_clusters_per_class=1,
                           weights=[0.95, 0.05], # Imbalance: 95% class 0,
                           5% class 1
                           class_sep=0.8, random_state=0)

# Apply Random Oversampling
ros = RandomOverSampler(random_state=42)
X_resampled, y_resampled = ros.fit_resample(X, y)

print("Original class distribution:", Counter(y))
print("Resampled class distribution:", Counter(y_resampled))
```

Expected Results

```
Original class distribution: Counter({0: 95, 1: 5})
Resampled class distribution: Counter({0: 95, 1: 95})
```

3. Synthetic Minority Oversampling Technique (SMOTE)

SMOTE is a sophisticated oversampling method that produces artificial instances for the minority group. SMOTE generates fresh points by interpolating between minority-class instances that are close in feature space, therefore enhancing the model's generalizability. Below is a sample code for SMOTE.

Listing 2-23. Using SMOTE to Generate Synthetic Data Points for the Minority Class

```
Import numpy as np
import pandas as pd
from imblearn.over_sampling import SMOTE
from sklearn.datasets import make_classification

# Sample imbalanced dataset (or use your own X, y)
X, y = make_classification(n_classes=2, class_sep=2,
                           weights=[0.1, 0.9], n_informative=3,
                           n_redundant=1, flip_y=0, n_features=20,
                           n_clusters_per_class=1, n_samples=1000,
                           random_state=42)

#Check class distribution before SMOTE
print(f"Original class distribution: {np.bincount(y)}")

# Initialize SMOTE
smote = SMOTE(random_state=42, sampling_strategy='auto')

# Apply SMOTE
try:
    X_res, y_res = smote.fit_resample(X, y)

    # Check new class distribution
    print(f"Resampled class distribution: {np.bincount(y_res)}")

    # Verify shapes
    print(f"X_res shape: {X_res.shape}, y_res shape: {y_res.shape}")

except Exception as e:
    print(f"Error occurred: {strI}")
```

Expected Results

```
Original class distribution: [100 900]
Resampled class distribution: [900 900]
X_res shape: (1800, 20), y_res shape: (1800,)
```

Algorithm-Level Techniques for Handling Imbalanced Data

In addition to manipulating the dataset, specific modifications can be made to the algorithms themselves to handle class imbalance.[19] In this section, we will talk about some algorithm-level techniques that deal with imbalanced data.

1. Cost-Sensitive Learning

Cost-sensitive learning assigns a higher misclassification cost to the minority class, and it causes the model to pay more attention to those instances. It adds weight to minor classes, which penalize errors on minority-class instances more than errors on majority-class instances. Most classification algorithms, such as decision trees, logistic regression, and support vector machines, can incorporate class weights to handle imbalance. The sample code below shows how to add weight to the minor class.

Listing 2-24. Applying Class Weights to a Logistic Regression Model

```
Import numpy as np
from sklearn.linear_model import LogisticRegression

# ---- Minimal sample data----
# 200 samples, 5 features
X = np.random.rand(200, 5)

# Imbalanced binary target: mostly 0s, few 1s
y = np.random.choice([0, 1], size=200, p=[0.9, 0.1])

# ---- Cost-sensitive Logistic Regression ----
model = LogisticRegression(class_weight='balanced')
model.fit(X, y)

print("Model coefficients:", model.coef_)
```

Expected Results

```
Model coefficients: [[-0.27516087  1.34217189 -0.40533604 -0.64006982
0.39983472]]
```

2. Ensemble Methods

Ensemble methods such as Balanced Random Forest and Easy Ensemble are designed to deal with imbalanced data. These methods use bootstrap sampling and model aggregation to enhance the performance of the minority class.

Balanced Random Forest

A modification of the standard Random Forest where each tree is trained on a bootstrap sample that has been randomly undersampled to balance the classes. This helps prevent the forest from being biased toward the majority class.

Easy Ensemble

This method repeatedly creates balanced subsets of the data via random undersampling, trains a weak learner (often AdaBoost) on each subset, and then combines their outputs. It is effective in improving minority-class performance while reducing overfitting.

Evaluation Metrics for Imbalanced Data

In working with imbalanced classes, precision, recall, and F1 score are more insightful than accuracy. Precision quantifies the number of true positives among anticipated ones, whereas Recall gauges the number of correctly predicted actual positives. The F1 score gives a middle ground between accuracy and recall.[19]

ROC Curve and AUC: The Receiver Operating Characteristic (ROC) curve plots the True Positive Rate (Recall) against the False Positive Rate across different classification thresholds. The Area Under the Curve (AUC) summarizes the model's ability to distinguish between classes. A higher AUC suggests better overall separability between the minority and majority classes.

Confusion Matrix: By showing true positives, false positives, true negatives, and false negatives, it offers a full view of model performance. This aids in evaluating the model's accuracy in predicting minority and majority groups.

Listing 2-25. Generating a Confusion Matrix with `scikit-learn`

```
From sklearn.metrics import confusion_matrix
import numpy as np

# Sample true and predicted labels
y_true = np.array([0, 1, 0, 1, 1, 0, 1])
y_pred = np.array([0, 1, 0, 0, 1, 0, 1])

# Generate confusion matrix
cm = confusion_matrix(y_true, y_pred)
print(cm)
```

Expected Results

```
[[3 0]
 [1 3]]
```

Scaling Data Pipelines for Performance

Machine learning pipeline projects fall into two different parts: data and model. First, we will focus on scaling data pipelines with techniques like distributed processing of data. This technique helps to use cloud resources and specialized frameworks for better performance.

Distributed Data Processing

When working with a large volume of data that cannot fit into a single machine, distributed data processing is a good solution. Big data frameworks like Apache Spark and Hadoop allow for parallel processing of data across a cluster of machines and speed up processing.[10] The code below shows a sample for processing a large dataset using PySpark.

Listing 2-26. Using PySpark to Process a Large Dataset

```
From pyspark.sql import SparkSession

# Create a Spark session
spark = SparkSession.builder.appName("DataPipelineScaling").getOrCreate()

# Load a CSV file into a DataFrame
df = spark.read.csv("large_dataset.csv", header=True, inferSchema=True)

# Perform transformations
df_filtered = df.filter(df['value'] > 100)

# Show the results
df_filtered.show()
```

Leveraging Cloud Infrastructure

Cloud platforms such as AWS, Google Cloud Platform (GCP), and Azure make it simple to scale the data pipelines. These platforms provide managed services that simplify data processing, orchestration, and automation, enabling massive datasets to be managed and scaled. Table 2-4 compares the data processing services of AWS, GCP, and Azure.

The serverless services presented in this section offer data processing solutions that help data engineers and scientists focus on more important topics instead of infrastructure. AWS Glue is very good at data preparation and transformation and works perfectly with other AWS services. Google Dataflow works well in both stream and batch processing and is accountable for its scalability and performance. Azure Data Factory is another perfect option for integration, allowing effective orchestration of data from different sources. Table 2-4 shows a summary of these and similar services.

***Table 2-4.** Summary of Data Processing Services in Cloud Platforms*

Platform	Service	Description	Key Features	Use Cases
AWS	AWS Glue	Serverless ETL for data preparation and transformation.	Integrates with S3, Redshift; Spark-based	Data lakes, analytics, ETL workflows
GCP	Dataflow	Unified stream and batch data processing.	Apache Beam SDK; autoscaling	Streaming, batch processing, pipelines
Azure	Data Factory	Cloud ETL for data orchestration and movement.	Hybrid integration; synapse integration	Multi-source data pipelines, orchestration
AWS	EMR	Managed Hadoop and Spark for big data processing.	Spark, Hive, Presto; scalable clusters	Big data analytics, genomic data
GCP	BigQuery	Serverless data warehouse for real-time analytics.	SQL-based; built-in machine learning	Large-scale analytics, BI
Azure	Synapse Analytics	Unified data warehousing and big data analytics.	T-SQL; Power BI integration	Enterprise analytics, BI workflows
AWS	Lambda	Event-driven serverless computing for ETL tasks.	Low latency; event-driven	Real-time transformation, event handling
GCP	Pub/Sub	Messaging service for event-driven systems.	Real-time, low-latency delivery	Streaming, event processing
Azure	Stream Analytics	Serverless analytics for IoT and real-time data.	SQL queries; IoT Hub integration	IoT analytics, telemetry processing
AWS	Kinesis	Real-time data streaming platform.	Scalable; AWS service integration	Logs, IoT stream processing
GCP	Dataproc	Managed Spark and Hadoop for batch and stream jobs.	Fast cluster setup; GCS integration	Machine learning, data transformation
Azure	HDInsight	Managed an open source big data analytics service.	Supports Spark, Hadoop, Kafka	Big data, large-scale batch processing

Debugging Data Pipeline Issues

A robust machine learning system relies on well-processed, high-quality data. As data pipelines are responsible for collecting, transforming, and feeding data into ML models, any problem within these pipelines can result in performance degradation, model failures, and incorrect predictions. Despite traditional software bugs, data pipeline errors can be silent, producing misleading results. This makes debugging more complicated.[43] This section focuses on common data pipeline issues, methods for identifying and debugging them, and the best practices for increasing data quality in ML workflows.

Understanding Data Pipeline Issues

Data pipelines include several stages. These stages involve raw data ingestion to feature extraction and transformation. Errors can happen at any stage, making models learn from corrupted, incomplete, or biased data. There are various problems with data issues. One is that datasets may have missing values due to network issues, sensor failures, incomplete extractions, or other reasons. Or they can have corrupted data (e.g., mixed data types, incorrect encodings), which can disrupt feature processing. Another problem is that data distributions can alter over time, which makes models less effective in practical applications. In Chapter 6, in the monitoring part, you got familiar with two concepts: data drift, which refers to shifts in input features, and concept drift, which happens when the relationship between inputs and outputs changes. Another issue is feature leakage. Feature leakage happens when a feature (unintentionally) contains future information. This leads to over-optimistic performance during training and poor generalization in production. Another problem is data duplication and skew. Duplicate records affect the importance of certain examples, which results in biased models. Additionally, train-test skew happens when training and test data distributions are different. This makes performance estimation inaccurate. Data from different sources usually needs joins and aggregations, which may introduce missing values, misaligned timestamps, or duplicate entries. Did you know that data can suffer from so many issues? If that was too much information, I suggest reading the above part twice! We also discuss the above issues in more detail in this section.

Identifying Data Pipeline Issues

Identifying errors in ML data pipelines needs a mix of statistical analysis, automated validation, and visualization techniques. In this section, we will talk about effective debugging approaches.

Checking for Missing and Corrupt Data

One of the easiest ways to debug data pipeline problems is to inspect missing values and properly handle them. As we have discussed about missing data in this chapter, here we just quickly review that with an example, and we will not discuss it further. The following code checks for unusual data types and missing values.

Listing 2-27. Checking Unusual Data Types and Missing Values

```
import pandas as pd

# Load dataset
df = pd.read_csv("data.csv")

# Check for missing values
missing_values = df.isnull().sum()
print("Missing Values:\n", missing_values[missing_values > 0])

# Check for unexpected data types
print("\nData Types:\n", df.dtypes)
```

If there are missing values, you can handle them by using imputation techniques (e.g., mean, mode, median, or predictive modeling). For categorical data, a popular strategy is filling missing values with the most frequent category. The below code fills values with the mean and categorical data with the mode.

Listing 2-28. Imputing Missed Data

```
df.fillna(df.mean(), inplace=True)  # Impute numerical columns with mean
df.fillna(df.mode().iloc[0], inplace=True)  # Impute categorical columns
with mode
```

Detecting Data Drift and Concept Drift

Monitoring data drift is critical for ensuring that models keep on performing well in production.[44] The following code compares feature distributions between incoming (production) and training data using the Kolmogorov-Smirnov (KS) test. KS is a statistical test for distribution differences.

Listing 2-29. Comparing Feature Distribution Between Training and Production Data

```
from scipy.stats import ks_2samp
import numpy as np

# Load training and production datasets
train_data = pd.read_csv("train_data.csv")
prod_data = pd.read_csv("prod_data.csv")

# Compare feature distributions
for column in train_data.columns:
    stat, p_value = ks_2samp(train_data[column].dropna(), prod_
    data[column].dropna())
    if p_value < 0.05:
        print(f"Feature {column} has changed significantly
        (p={p_value:.5f})")
```

A low p-value shows that the feature distribution has considerably shifted, probably data drift. For identifying concept drift, monitoring changes in model prediction distributions over time is helpful. If the model's output distribution changes without corresponding changes in input features, concept drift is probable. The code below shows how to find concept drift.

Listing 2-30. Comparing Model Predictions Over Time

```
import matplotlib.pyplot as plt

plt.hist(train_data["predictions"], alpha=0.5, label="Training")
plt.hist(prod_data["predictions"], alpha=0.5, label="Production")
plt.legend()
plt.xlabel("Prediction Scores")
```

```
plt.ylabel("Frequency")
plt.title("Concept Drift Detection")
plt.show()
```

Many sample codes in this book are just going to show you how to write your own code for a specific purpose. You should learn from this code and, in your projects, include real data.

Identifying Feature Leakage

Feature leakage happens when data that is not available at inference time is used during training. This leads to overly optimistic model performance. To clarify better, assume we have a model that is going to predict if an employee is late or not, predicting "Islate". While we train the model, the training data has the field "minutesLate"! It may look funny, but this can happen in real projects too, and detecting related fields may not be as easy as this example. One popular method for detecting leakage is by checking feature importance. The code below checks for feature leakage.

Listing 2-31. Sample Code for Checking for Feature Leakage

```
from sklearn.ensemble import RandomForestClassifier
from sklearn.model_selection import train_test_split

# Load dataset
df = pd.read_csv("data.csv")
X = df.drop(columns=["target"])  # Features
y = df["target"]  # Target variable

# Train a quick feature importance model
X_train, X_test, y_train, y_test = train_test_split(X, y, test_size=0.3,
random_state=42)
model = RandomForestClassifier()
model.fit(X_train, y_train)

# Check feature importance
importances = pd.Series(model.feature_importances_, index=X.columns).sort_
values(ascending=False)
print("Feature Importance:\n", importances)

# High importance for unexpected features might indicate leakage
```

If you see a feature with future information, it may show leakage and should be removed from the training data.

Detecting Data Duplication and Skew

Duplicate records can bias models toward the most frequently occurring samples. Checking for duplicates is a simple but required debugging step.

Listing 2-32. Sample Code for Detecting Duplicates

```
# Check for duplicate rows
duplicate_rows = df[df.duplicated()]
print("Number of duplicate records:", duplicate_rows.shape[0])
```

The good thing about Python is that it has useful libraries for data analysis, such as finding null values, duplicates, and so on. You can see how we find duplicates; we just need one line of Python code. Train-test skew happens when the distribution of features in the training and test datasets significantly differs. The following code checks for differences in feature means.

Listing 2-33. Checking Differences in Feature Means

```
train_mean = train_data.mean()
test_mean = test_data.mean()

diff_percentage = ((train_mean - test_mean) / train_mean) * 100
print("Feature distribution differences:\n", diff_percentage)
```

If some features show high percentage differences, they may need resampling, normalization, or better feature selection.

Best Practices for Preventing Data Pipeline Issues

I remember that when I was a kid, there was an advertisement about preventing and curing medical diseases. For prevention, it was showing an open way, and for cure, it was showing a door with guards. I have the same idea about the ML project, and I believe preventing data pipeline failures is better, easier, and more effective than debugging them. In this part, we discuss the best practices for minimizing errors and maintaining data integrity. Firstly, use tools like **Great Expectations** for integrity checks and enforcing schema validation. Do not forget the usefulness of tools and the effect of automation in your project. Using these tools or in your code, you can define acceptable data types, value ranges, and missing data thresholds. Then, by monitoring and logging data quality, you can track distribution changes, missing values, and anomalies in data. Also, I suggest setting up automated alerts when drift is identified. With this procedure, finding issues would be very easy. Also, you should use version control for features and data. With version control, you can track dataset versions and keep metadata logs for preprocessing and transformation steps. Another good practice is to add automated data checks before retraining models. And finally, ensure data pipeline changes do not create any inconsistencies.

Project: Titanic Survival Prediction Project

The Titanic was a tragic disaster in which many passengers lost their lives. In the project of this chapter, we are going to use the public data of the Titanic survival to practice our learning in this chapter.

Project Objective

In this project, you would build a robust machine learning pipeline using the Titanic dataset. We are going to predict whether a passenger has survived the tragic sinking of the Titanic or not. By doing this project, you will gain practical knowledge of the most popular data science tools, frameworks, and libraries, and you will deepen your insight into the full machine learning workflow. I should mention that all the codes of this project and other projects and examples are available on GitHub.

This project takes the Titanic dataset, a well-known benchmark for data science and machine learning projects, as a reference to predict passengers' chances of survival. Here, you are going to perform the complete work, which includes data collection, data exploration, missing and duplicate data handling, and feature engineering for predictive performance improvement. Finally, you will build a machine learning model and evaluate it using a Random Forest classifier. The project would be followed by a discussion of various options for scaling the pipeline using distributed computing frameworks or cloud platforms.

Project Description

This project consists of seven steps as follows:

1. **Importing Required Libraries**: The first step starts by loading all necessary libraries for modeling, visualization, and data manipulation to set the scene.
2. **Data Collection**: Download or load the Titanic dataset and examine it. In this step, a first visual of missing values as well as an overall summary would be produced.
3. **Data Cleaning**: Includes addressing missing values, eliminating irrelevant columns and duplicates, and correcting data discrepancies.
4. **Feature Engineering**: Create and transform features to improve the predictive power of the model.
5. **Scaling and Preprocessing**: This step creates pipelines for the preprocessing of numerical and categorical data to ensure consistency across model training.
6. **Build and Train the Model**: In this step, preprocessing is combined with a Random Forest classifier to train the model and assess its performance.
7. **Performance Pipeline Scaled**: The final step is to consider techniques for scaling the model using cloud systems or distributed frameworks.

Step 1: Import Necessary Libraries

In the first step, we are importing the required libraries for data manipulation, machine learning, and visual analysis. These libraries have three main goals and can be categorized as follows:

1. **Data Processing**: `NumPy` and `pandas`
2. **Model Development and Evaluation**: `scikitlearn`
3. **Visualizing**: `matplotlib` and `seaborn`

These libraries offer everything you need to create a full machine learning pipeline. The code below loads data.

Listing 2-34. Step 1: Loading Libraries

```
import pandas as pd
import numpy as np
from sklearn.model_selection import train_test_split
from sklearn.preprocessing import StandardScaler, OneHotEncoder
from sklearn.compose import ColumnTransformer
from sklearn.pipeline import Pipeline
from sklearn.impute import SimpleImputer
from sklearn.ensemble import RandomForestClassifier
from sklearn.metrics import accuracy_score, classification_report,
confusion_matrix
import matplotlib.pyplot as plt
import seaborn as sns
```

Expected Results

Apart from the successful importing of libraries, this step does not produce any results. The lack of errors shows all needed modules are present for the rest of the project.

Step 2: Data Collection

By downloading the Titanic dataset from a public URL, this step takes care of data collection. Data manipulation is continued by importing the dataset into a pandas DataFrame. Printing the first several rows and fundamental data on the dataset offers a first look; a heatmap is generated, which shows missing values and illustrates where additional cleaning might be needed. Earlier in this chapter, we discussed how we can import data from different sources, such as an API. In this part, we will practice pulling data from public APIs. Note that many public APIs provide free access to datasets. The code below loads data.The Titanic dataset is loaded straight from GitHub via pandas. While `data.info()` offers information on the number of entries, data types, and nonnull counts—vital for understanding data integrity—using `data.head()` gives a quick overview of the dataset. The Seaborn-generated heatmap (Figure 2-4) shows the columns with missing values, therefore directing the data cleaning.

Listing 2-35. Step 2: Data Collection

```
url = "https://raw.githubusercontent.com/datasciencedojo/datasets/master/
titanic.csv"
data = pd.read_csv(url)

# Display the first few rows of the dataset
print("Data Overview:")
print(data.head())

# Display basic information about the dataset
print("\nDataset Information:")
print(data.info())

# Visualize missing values using a heatmap
plt.figure(figsize=(10, 6))
sns.heatmap(data.isnull(), cbar=False, cmap='viridis')
plt.title('Missing Values Heatmap')
plt.show()
```

Expected Results

The console shows the first several rows of the dataset, with an overview of its structure (number of entries, nonnull counts, data types). Visually, the heatmap shows those columns with absent data and helps you to plan for data cleaning. This step may not look necessary for building a machine learning project; however, I strongly recommend having a similar process in your ML projects.

```
Data Overview:
   PassengerId  Survived  Pclass  \
0            1         0       3
1            2         1       1
2            3         1       3
3            4         1       1
4            5         0       3
```

```
                                                Name     Sex   Age  SibSp \
0                            Braund, Mr. Owen Harris    male  22.0      1
1                            Cumings, Mrs. John Bradley (Florence Briggs
Th...  female  38.0      1
2                             Heikkinen, Miss. Laina  female  26.0      0
3       Futrelle, Mrs. Jacques Heath (Lily May Peel)  female  35.0      1
4                           Allen, Mr. William Henry    male  35.0      0

   Parch            Ticket     Fare Cabin Embarked
0      0         A/5 21171   7.2500   NaN        S
1      0          PC 17599  71.2833   C85        C
2      0  STON/O2. 3101282   7.9250   NaN        S
3      0            113803  53.1000  C123        S
4      0            373450   8.0500   NaN        S

Dataset Information:
<class 'pandas.core.frame.DataFrame'>
RangeIndex: 891 entries, 0 to 890
Data columns (total 12 columns):
 #   Column       Non-Null Count  Dtype
---  ------       --------------  -----
 0   PassengerId  891 non-null    int64
 1   Survived     891 non-null    int64
 2   Pclass       891 non-null    int64
 3   Name         891 non-null    object
 4   Sex          891 non-null    object
 5   Age          714 non-null    float64
 6   SibSp        891 non-null    int64
 7   Parch        891 non-null    int64
 8   Ticket       891 non-null    object
 9   Fare         891 non-null    float64
 10  Cabin        204 non-null    object
 11  Embarked     889 non-null    object
dtypes: float64(2), int64(5), object(5)
memory usage: 83.7+ KB
None
```

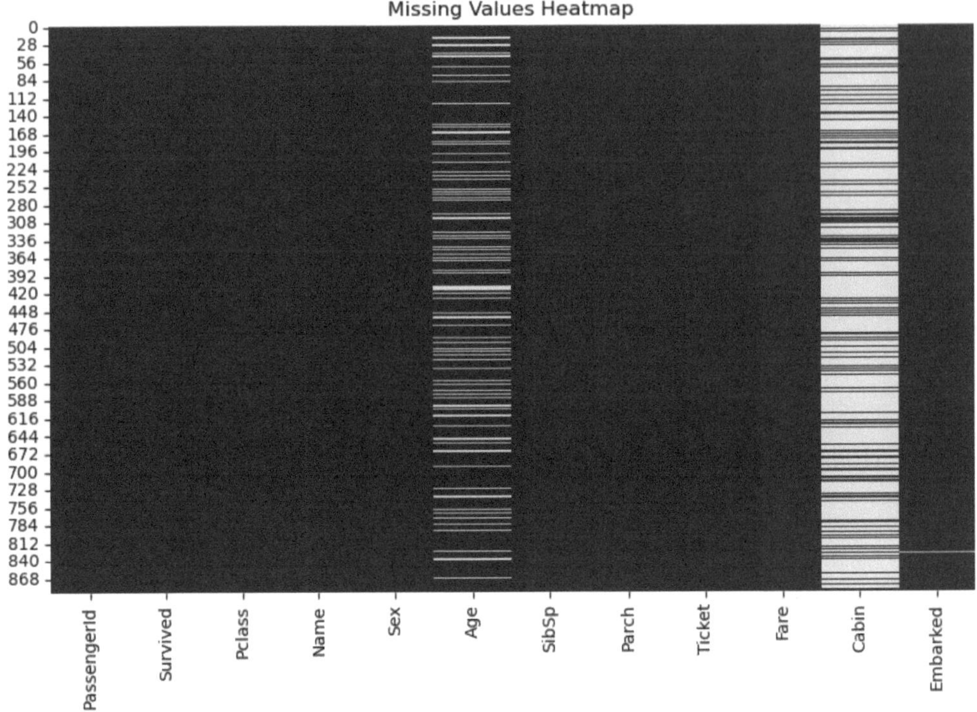

Figure 2-4. *Heatmap Chart for Titanic Passenger Data*

Step 3: Data Cleaning

Cleaning the data means dealing with missing values, getting rid of superfluous columns, and deleting repeated entries. Before feature engineering and modeling can be started, this process is required to verify that the dataset is consistent and dependable. Earlier in this chapter, we discussed data cleaning in detail. In this step, you will apply the techniques we covered. The code below performs the data cleaning.

This part starts with the counting of missed entries for every column. Due to their high missing proportion or low predictive value, unimportant columns like "Name", "Ticket", and "Cabin" are eliminated. Remaining columns with missing numeric values in "Age" are filled with the median. To decrease the influence of outliers, missing categorical entries in "Embarked" are imputed with the mode. Data integrity is guaranteed by the detection and deletion of duplicate rows. A count plot is shown in Figure 2-5 that helps to see the distribution of the target variable to ascertain if there are class discrepancies.

Listing 2-36. Step 3: Data Cleaning

```
# Check for missing values before cleaning
print("\nMissing Values Before Cleaning:")
print(data.isnull().sum())

# Drop irrelevant columns (e.g., 'Name', 'Ticket', 'Cabin')
# These columns are either too specific or have too many missing values
data_cleaned = data.drop(columns=['Name', 'Ticket', 'Cabin'],
errors='ignore')

# Impute missing values for numerical columns (e.g., 'Age') with the median
data_cleaned['Age'].fillna(data_cleaned['Age'].median(), inplace=True)

# Impute missing values for categorical columns (e.g., 'Embarked') with
the mode
data_cleaned['Embarked'].fillna(data_cleaned['Embarked'].mode()[0],
inplace=True)

# Drop rows with missing values in the target column (if any)
data_cleaned.dropna(subset=['Survived'], inplace=True)

# Verify cleaning results
print("\nMissing Values After Cleaning:")
print(data_cleaned.isnull().sum())

# Check for duplicate rows and remove them if any
print(f"\nNumber of Duplicate Rows Before Removal: {data_cleaned.
duplicated().sum()}")
data_cleaned.drop_duplicates(inplace=True)
print(f"Number of Duplicate Rows After Removal: {data_cleaned.duplicated().
sum()}")

# Visualize the distribution of the target variable ('Survived')
plt.figure(figsize=(6, 4))
sns.countplot(x='Survived', data=data_cleaned)
plt.title('Distribution of Survived')
plt.show()
```

Expected Results

After running the code of this step, you should notice a reduction or removal of missing values and duplicates in the printed data. An important measure for grasping any class imbalances affecting model performance, the `countplot` shows the survival results.

```
Missing Values Before Cleaning:
PassengerId      0
Survived         0
Pclass           0
Name             0
Sex              0
Age            177
SibSp            0
Parch            0
Ticket           0
Fare             0
Cabin          687
Embarked         2
dtype: int64

Missing Values After Cleaning:
PassengerId    0
Survived       0
Pclass         0
Sex            0
Age            0
SibSp          0
Parch          0
Fare           0
Embarked       0
dtype: int64

Number of Duplicate Rows Before Removal: 0
Number of Duplicate Rows After Removal: 0
```

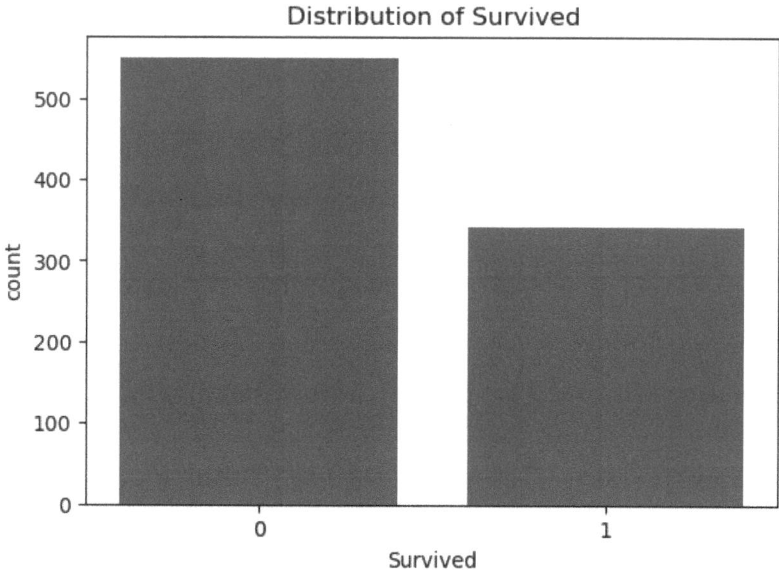

Figure 2-5. *A Count Plot to See the Data Distribution*

Step 4: Feature Engineering

As we discussed earlier, feature engineering includes making new variables (features) from existing features to improve the performance of the model. In this step, you will create some additional features that get the important relationships in the data, such as family size and age groups, for example. Here in the code, new features are explained in comments. The code below shows feature engineering.

This step included several transformations:

`FamilySize`**:** Can be computed by adding the number of family members (parents/children). As well as do not forget to add one to count the passengers themselves.

`Sex Conversion`**:** This feature is converted from the categorical string (Male, Female) to a binary numeric variable.

`IsAlone`**:** A binary feature based on `FamilySize` signaling lone trips.

`AgeGroup`**:** To capture nonlinear relationships with life expectancy, the continuous "Age" variable is grouped into categories (Child, Teen, Adult, Senior).

The changed `DataFrame` is displayed after these transformations, and a `countplot`, Figure 2-6, shows the survival rates by age group.

Listing 2-37. Step 4: Feature Engineering

```
# Feature Extraction: Create a new feature 'FamilySize' from 'SibSp'
and 'Parch'
data_cleaned['FamilySize'] = data_cleaned['SibSp'] + data_
cleaned['Parch'] + 1

# Feature Transformation: Convert 'Sex' into binary values (0 for male, 1
for female)
data_cleaned['Sex'] = data_cleaned['Sex'].map({'male': 0, 'female': 1})

# Feature Creation: Create a new feature 'IsAlone' based on 'FamilySize'
data_cleaned['IsAlone'] = (data_cleaned['FamilySize'] == 1).astype(int)

# Feature Transformation: Binning 'Age' into categories (e.g., Child, Teen,
Adult, Senior)
bins = [0, 12, 18, 60, 100]
labels = ['Child', 'Teen', 'Adult', 'Senior']
data_cleaned['AgeGroup'] = pd.cut(data_cleaned['Age'], bins=bins,
labels=labels)

# Display the updated dataset
print("\nDataset After Feature Engineering:")
print(data_cleaned.head())

# Visualize the new 'AgeGroup' feature with respect to survival
plt.figure(figsize=(8, 5))
sns.countplot(x='AgeGroup', hue='Survived', data=data_cleaned)
plt.title('Survival Rate by Age Group')
plt.show()
```

Expected Results

Now the dataset is changed to a more suitable form for machine learning. The new printed dataset includes some new columns such as "`FamilySize`", "`IsAlone`", and "`AgeGroup`". The plot illustrates the distribution of survival outcomes by age group, revealing any trends or patterns in the data. Here, as it was just a simple project, we could easily guess some important features, and we included them in our data. However, in more complicated practical ways, talking with an expert is recommended. Again, I emphasize, talk with an expert before developing your real-world ML project.

```
Dataset After Feature Engineering:
   PassengerId  Survived  Pclass  Sex   Age  SibSp  Parch     Fare
Embarked  \
0            1         0       3    0  22.0      1      0   7.2500        S
1            2         1       1    1  38.0      1      0  71.2833        C
2            3         1       3    1  26.0      0      0   7.9250        S
3            4         1       1    1  35.0      1      0  53.1000        S
4            5         0       3    0  35.0      0      0   8.0500        S

   FamilySize  IsAlone AgeGroup
0           2        0    Adult
1           2        0    Adult
2           1        1    Adult
3           2        0    Adult
4           1        1    Adult
```

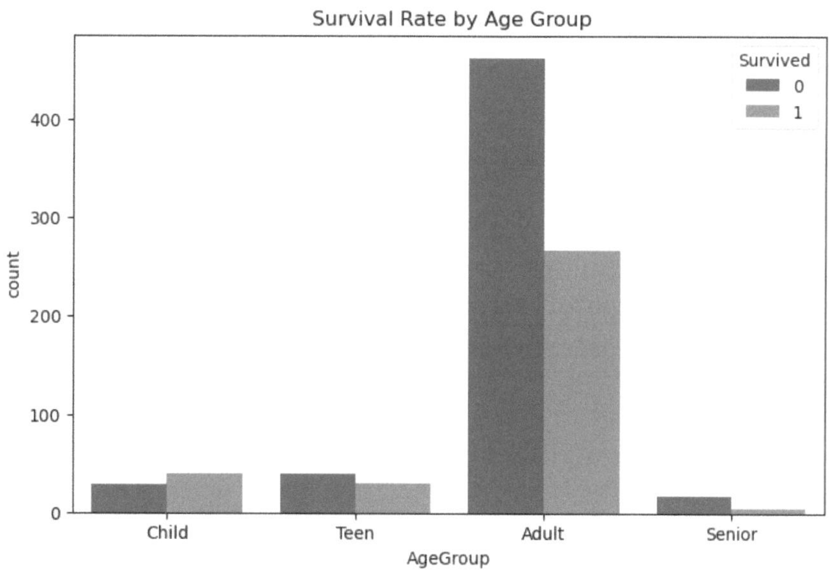

***Figure 2-6.** Age Group Distribution of Survival*

Step 5: Scaling and Preprocessing

The next step is to build a preprocessing pipeline to make the data ready for model training. This process guarantees that the data is in the right format for machine learning models by imputing, scaling, and one-hot encoding both for numerical and categorical attributes. The code below performs the scaling and preprocessing.

In this step, the features (X) and the target variable (y) are separated. The pipeline that imputes missing values with the median and scales the numerical characteristics in a standard way runs (“Age”, “Fare”, “FamilySize”). Another pipeline handles the categorical features (“Sex”, “Embarked”, “Pclass”, “AgeGroup”). This pipeline fills missing values with the most common entry and applies one-hot encoding to transform categories into binary vectors. A `ColumnTransformer` is used to combine these pipelines so that every feature set is handled correctly, guaranteeing a constant transformation for every future prediction and training.

Listing 2-38. Step 5: Scaling and Preprocessing

```
# Separate features and target variable
X = data_cleaned.drop(columns=['Survived'])
y = data_cleaned['Survived']

# Define numerical and categorical columns for preprocessing
numerical_cols = ['Age', 'Fare', 'FamilySize']
categorical_cols = ['Sex', 'Embarked', 'Pclass', 'AgeGroup']

# Preprocessing for numerical data: Impute missing values and scale
numerical_transformer = Pipeline(steps=[
    ('imputer', SimpleImputer(strategy='median')),
    ('scaler', StandardScaler())
])

# Preprocessing for categorical data: Impute missing values and one-
hot encode
categorical_transformer = Pipeline(steps=[
    ('imputer', SimpleImputer(strategy='most_frequent')),
    ('onehot', OneHotEncoder(handle_unknown='ignore'))
])
```

```
# Combine preprocessing steps using ColumnTransformer
preprocessor = ColumnTransformer(
    transformers=[
        ('num', numerical_transformer, numerical_cols),
        ('cat', categorical_transformer, categorical_cols)
    ]
)
```

Expected Results

This step establishes a modular preprocessing pipeline to be incorporated with the machine learning model. While no direct result is created here, you now have a powerful tool to handle both numerical and categorical data consistently. In this part, having no result is a good sign that the code has run without any errors. In all steps, you can change the codes and play with them to get more insight into how it works.

Step 6: Build and Train the Model

In this step, the `scikit-learn` Pipeline class links the preprocessing pipeline with a Random Forest classifier. Then, the merged pipeline is trained on the dataset, and the model is assessed with accuracy metrics, a thorough classification statement, and a confusion matrix. The code below builds and trains the model.

By joining the earlier defined preprocessing stages with a Random Forest classifier, this step sets up the whole machine learning pipeline. The dataset is divided into training and testing samples (80/20 split) to simplify the model evaluation of the unseen data. The training set is used to train the model, which then creates forecasts on the test. Here, we discuss the main component of HITL. The first item is Modset. Finally, the performance is evaluated by calculating the accuracy score, generating a classification report (which outlines precision, recall, and F1 scores), and graphing the outcomes with a confusion matrix heatmap. Figure 2-7 shows the confusion matrix.

Listing 2-39. Step 6: Build and Train the Model

```
model = Pipeline(steps=[
    ('preprocessor', preprocessor),
    ('classifier', RandomForestClassifier(random_state=42))
])
```

```
# Split the data into training and testing sets
X_train, X_test, y_train, y_test = train_test_split(X, y, test_size=0.2,
random_state=42)

# Train the model on the training data
model.fit(X_train, y_train)

# Make predictions on the test set
y_pred = model.predict(X_test)

# Evaluate the model's performance by calculating accuracy
accuracy = accuracy_score(y_test, y_pred)
print(f"\nModel Accuracy: {accuracy:.2f}")

# Print a detailed classification report
print("\nClassification Report:")
print(classification_report(y_test, y_pred))

# Plot the confusion matrix
plt.figure(figsize=(6, 4))
sns.heatmap(confusion_matrix(y_test, y_pred), annot=True, fmt='d',
cmap='Blues')
plt.title('Confusion Matrix')
plt.xlabel('Predicted')
plt.ylabel('Actual')
plt.show()
```

Expected Results

Once you run this code, you will see displayed metrics including a complete classification report and model accuracy (e.g., "`Model Accuracy: 0.83`"). The confusion matrix heatmap gives a visual breakdown of correct vs. incorrect forecasts.

```
Model Accuracy: 0.83

Classification Report:
              precision    recall  f1-score   support

           0       0.84      0.87      0.85       105
           1       0.80      0.77      0.79        74
```

```
    accuracy                           0.83       179
   macro avg       0.82      0.82      0.82       179
weighted avg       0.83      0.83      0.83       179
```

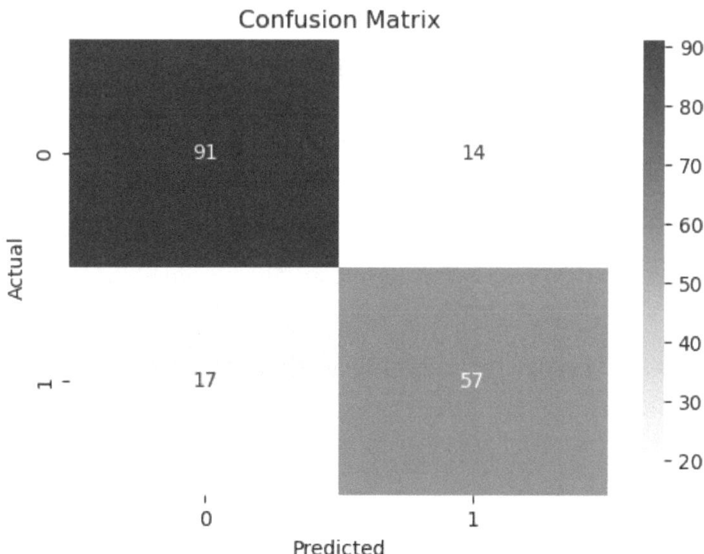

Figure 2-7. *Confusion Matrix*

Step 7: Scaling the Pipeline for Performance

For practical uses, machine learning pipelines usually have to handle big datasets. This last stage involves scaling the pipeline via cloud services or distributed computing frameworks. If you wish to investigate distributed data processing, the Dask optional example is provided. The below code scales the pipeline for performance.

This step offers ideas on how to scale the machine learning pipeline to process more data. Even though the Dask code is commented out, it illustrates how a pandas DataFrame may be converted to a Dask DataFrame to allow concurrent processing. Furthermore, the model can be scaled via cloud-based systems (e.g., AWS SageMaker or Google Cloud AI Platform). The print statement validates the view that the pipeline is organized to support scaling.

Listing 2-40. Step 7: Scaling the Pipeline for Performance

```
import dask.dataframe as dd
import pandas as pd

# Convert pandas DataFrame to Dask DataFrame for distributed processing
dask_data = dd.from_pandas(data_cleaned, npartitions=2)

# Perform operations on Dask DataFrame, filling numeric missing values
dask_data = dask_data.map_partitions(
  lambda df: df.fillna(df.median(numeric_only=True)),
  meta=data_cleaned
)

print("\nPipeline is ready for scaling with distributed frameworks or cloud
platforms!")
```

Expected Results

```
Pipeline is ready for scaling with distributed frameworks or cloud
platforms!
```

This step produces a confirmation message that the pipeline is good for scaling. Although no actual visual output is present here, you can see what scaling the machine learning pipelines means. You can use a similar approach for scaling your big projects. As machine learning models are very big, scaling is as important as developing machine learning models. We will talk more about scaling in Chapter 4.

Conclusion

In this project, you built an end-to-end machine learning pipeline for predicting Titanic survival. You started by collecting and cleaning the data, handling missing values and duplicates. Then you engineered new features, created a preprocessing pipeline for numerical and categorical data, and combined it with a Random Forest classifier. You also explored how to scale the pipeline using distributed computing or cloud platforms which is an important step after training and evaluating the model.

This project demonstrated the full machine learning workflow and prepared you to work with larger, more complex datasets in real-world applications.

Summary

This chapter covered the main steps in building effective data pipelines: data collection, cleaning, and feature engineering. We discussed best practices for transforming data, handling large datasets, and using distributed or cloud-based tools.

We also reviewed key evaluation metrics such as precision, recall, F1 score, and ROC-AUC and methods for dealing with imbalanced data.

The Titanic project helped you apply these concepts from data loading to model building and pipeline scaling. A well-designed data pipeline is essential for successful machine learning, and you are now ready to apply these skills to real problems. The next chapter will focus on scaling and optimizing machine learning models for business applications.

CHAPTER 3

Selecting and Optimizing Machine Learning Models

Designing sophisticated artificial intelligence systems depends critically on model selection, as it greatly affects the accuracy and generalizability of the ultimate product. Choosing the proper model that has a balance among complexity, interpretability, and computational efficiency is sometimes the make-or-break of a machine learning initiative. Choosing the proper model for a particular issue depends on several elements, including the data type, the field of the problem, the availability of computational resources, as well as the desired performance measures. From easy linear regression exercises to difficult image recognition problems, each of which might call for a different approach to machine learning. Model selection also involves evaluating several trade-offs between overfitting and underfitting. In Chapter 1, we had a brief discussion about various machine learning models. Here, we will discuss several model selection methods and the advantages and disadvantages of different model types, and we will offer suggestions on selecting the most appropriate model for various machine learning applications.

Selecting a Suitable Model for the ML Project

In Chapter 2, we talked about how to prepare data for machine learning model development. Now the data is ready, and we are going to select a model and train it. The problem is that there are many different models, and selecting the most suitable one should be done wisely. Because the selected model will affect the performance and final result of the project. Here, as the first step, we talk about different types of ML models and their cons and pros. Based on that, we can select the most suitable model for a project.

M. R. Mahdiani, *Mastering Machine Learning Architecture and Solutions*,
https://doi.org/10.1007/979-8-8688-2527-9_3

Types of Models

Depending on the nature of the problem and the availability, different types of models are employed in machine learning. The primary categories include

Linear Models: Ideal for situations in which there is a linear correlation between the target variable and characteristics. Examples are linear regression and logistic regression.

Tree-Based Models: These include decision trees, Random Forests, and Gradient Boosting Machines, which are well-suited for handling nonlinear relationships and complex datasets.

Support Vector Machines (SVM): Particularly useful for both classification and regression tasks when the data is not linearly separable.

Deep learning models like convolutional neural networks (CNNs) and recurrent neural networks (RNNs) are very good for sophisticated tasks with images, sequences, and enormous sets.

Ensemble Models: Techniques such as Random Forests, Gradient Boosting, and Stacking combine many models to increase general accuracy and strength. Table 3-1 shows how various types of models differ in terms of their strengths, weaknesses, and common use cases. You can select the suitable models based on the strengths, weaknesses, and common use cases that are listed in this table. I encourage you to be creative and sometimes think outside the box in model selection! In the next part, we will discuss more about the criteria of model selection.

Table 3-1. *Comparison of Different Types of Models*

Model Type	Strengths	Weaknesses	Common Use Cases
Linear Models	Simple, interpretable, fast training	Limited to linear relationships	Regression, binary classification
Tree-Based Models	Handles nonlinearity, interpretable	Prone to overfitting without tuning	Classification, regression
SVM	Effective in high-dimensional spaces	Requires careful tuning of hyperparameters	Classification, outlier detection
Neural Networks	Handles complex data, state-of-the-art accuracy	Requires large datasets, computationally expensive	Image, NLP, time-series
Ensemble Models	Reduces overfitting, improves accuracy	Complex to implement and interpret	Kaggle competitions, high-stakes predictions

Model Selection Criteria

Selecting the right model depends on the evaluation of some main factors. In this section, we will discuss these criteria.

Nature of Data: Different models perform better depending on the data type. Linear models are best for numerical data with linear relationships, but neural networks are outstanding at dealing with unstructured data like text and images.

Complex vs. Simple Models: Linear regression and decision trees are easy but more explainable, which is critical for particular industries such as healthcare and finance. However, more sophisticated models such as deep neural networks could offer improved performance, but at the cost of interpretability.

Training Time and Computer Resources: Models such as neural networks need much training time and huge computing resources; however, if you do not have data centers or supercomputers with many GPUs, you can implement models such as logistic regression and decision trees that are less computationally expensive and fast to train. A trade-off is required depending on the available resources.

Performance Criteria: The choice of model depends on the metrics used to assess it. Depending on the demands of the problem, metrics including accuracy, precision, recall, F1 score, and AUC ROC can help with model selection. For instance, in an imbalanced dataset, one might like models with high recall.

Overfitting and Underfitting: Understanding how well a model generalizes to unseen data is important with regard to overfitting and underfitting trends. If not correctly regularized, models like decision trees tend to overfit; simpler models could underfit the information. By way of cross-validation, one can establish whether a model is overfitting or underfitting.

Cross-Validation for Model Selection

Cross-validation is a strong strategy for evaluating model performance and avoiding overfitting. In cross-validation, data is divided into two sets: a training set and a test set. The model is trained using the training set and then is tested over the test set, which is not used in model training. I highly recommend considering cross-validation in all your projects. Assume that the model is trained and tested k times in k-fold cross-validation. The data is split into k subsets, and each fold serves as the validation set once, while the remaining folds are used for model training. By averaging the result across overall folds, this approach helps to give a more precise estimation of model performance.

Another common approach is leave-one-out cross-validation (LOOCV), where each datum point acts as the validation set once. LOOCV is computationally expensive, but it offers an impartial appraisal of model quality, particularly when the dataset is tiny.[12]

Automated Model Selection

Automated tools like Grid Search and Randomized Search help in locating the optimal model and hyperparameters. Grid Search is a thorough approach that considers every conceivable parameter combination; Randomized Search is more effective since it randomly samples a set number of parameter combinations; therefore, it is best for more extensive search areas. Yet another sophisticated technique, Bayesian Optimization, uses probabilistic models to find the most promising hyperparameters, thus saving time relative to exhaustive search methods. By repeatedly refining the search based on the most probable results from areas of parameter space, Bayesian Optimization helps to increase convergence and make sensible use of computer resources.

`Scikitlearn` library offers user-friendly code for automatic model selection and hyperparameter optimization. Below is a sample code that shows how to apply Grid Search with `scikitlearn` to choose the ideal model for a classification problem.

Listing 3-1. Using Grid Search to Select the Best Model for a Classification Problem

```
# Grid search example
from sklearn.model_selection import GridSearchCV
from sklearn.ensemble import RandomForestClassifier
from sklearn.svm import SVC
from sklearn.datasets import load_iris

# Load sample dataset
data = load_iris()
X, y = data.data, data.target

# Define models and parameters
models = {
    'RandomForest': RandomForestClassifier(),
    'SVM': SVC()
}
```

```
params = {
    'RandomForest': {'n_estimators': [10, 50, 100]},
    'SVM': {'C': [0.1, 1, 10], 'kernel': ['linear', 'rbf']}
}

# Perform Grid Search
for model_name in models:
    grid_search = GridSearchCV(models[model_name], params[model_
    name], cv=5)
    grid_search.fit(X, y)
    print(f"Best parameters for {model_name}: {grid_search.best_params_}")
```

Excepted Results

```
Best parameters for RandomForest: {'n_estimators': 50}
Best parameters for SVM: {'C': 1, 'kernel': 'linear'}
```

Model selection is very important in machine learning; hence, a balance among performance, interpretability, and computational efficiency is needed. Applying systematic techniques like cross-validation and automated hyperparameter tuning, along with knowing the pros and cons of various models, will increase the quality of the machine learning solutions. By thoroughly considering the criteria and trying several models, you may choose the most appropriate model to fit your problem goals. After selecting the model, we need to optimize the hyperparameters of the model to ensure the highest performance. In the next section, we will talk about hyperparameter optimization.

Hyperparameter Optimization

The machine learning process depends on a great deal on hyperparameter optimization, so hyperparameter optimization is a crucial stage. Regularization parameters, learning rate, number of hidden layers in a neural network, and number of estimators in a Random Forest are among the hyperparameters. Be aware that the parameters that are set during the model training, like weight and bias, are not considered as hyperparameters. Hyperparameter proper tuning makes sure the model generalizes well to fresh, unseen data, in addition to being nicely trained.[13,14]

In this part, we will go over techniques including Grid Search, Randomized Search, Bayesian Optimization, hyperband, and genetic algorithms.

Hyperparameter Types

There are two main sets of hyperparameters:

Model-Specific Hyperparameters: These are configuration options affecting a given type of model's learning and formation. For example, the number of layers in a neural network or the count of trees in a Random Forest.

Training Hyperparameters: These determine the training process itself, including learning rate, batch size, and number of epochs. Choosing the correct hyperparameters can seriously affect the model's learning performance.

Now that we know different types of hyperparameters, we can proceed with techniques for selecting the best hyperparameters (optimized values).

Techniques for Hyperparameter Optimization

Hyperparameter optimization is a vital stage in the construction of machine learning models because it directly affects the performance, efficiency, and generalization abilities of the algorithms. Hyperparameters, which include learning rates, regularization coefficients, and the number of layers in a neural network, are configuration settings outside the model. They need an early setting; unlike model parameters, they are not learned during training. Choosing the perfect hyperparameter values could distinguish a model that barely performs from one that meets state-of-the-art standards.[15,16] There are various techniques to optimize hyperparameters. In the following, we will discuss the most important ones.

Grid Search

In Grid Search, instead of a predefined set of hyperparameters, a brute force approach is used. It means all combinations of hyperparameters are tested, and the combination yielding the best performance is chosen. Grid Search is effective when the hyperparameter search space is small, since it thoroughly analyzes all potential combinations. The code below shows a sample code of Grid Search in optimizing support vector machine learning.

Listing 3-2. Using Grid Search to Optimize a Support Vector Machine (SVM) Model

```
from sklearn.model_selection import GridSearchCV
from sklearn.svm import SVC

# Define the model and hyperparameters
grid_params = {
    'C': [0.1, 1, 10],
    'kernel': ['linear', 'rbf']
}

svc = SVC()
grid_search = GridSearchCV(svc, grid_params, cv=5)

grid_search.fit(X, y)
print(f"Best parameters: {grid_search.best_params_}")
```

Expected Results

```
Best parameters: {'C': 1, 'kernel': 'linear'}
```

Limitations: Grid Search can become computationally expensive in the case that there are many hyperparameters, especially if each has a wide range of possible values. The exponential growth of possible combinations makes it difficult for bigger models.

Randomized Search

By randomly sampling a fixed number of hyperparameter combinations from the search space, Randomized Search helps to overcome the constraints of Grid Search. Dealing with big search spaces, this method is much more effective since it doesn't try every potential combination. The code below shows how to use a Randomized Search to optimize the Random Forest model.

Listing 3-3. Using Randomized Search to Optimize a Random Forest Model

```
from sklearn.model_selection import RandomizedSearchCV, train_test_split
from sklearn.ensemble import RandomForestClassifier
from sklearn.datasets import make_classification  # For sample data
```

```
# Generate some sample data (replace with your actual data)
X, y = make_classification(n_samples=100, n_features=20, random_state=42)

# Split data into training and testing sets
X_train, X_test, y_train, y_test = train_test_split(X, y, test_size=0.2,
random_state=42)

# Define the model and hyperparameters (CORRECTED)
rand_params = {
    'n_estimators': [10, 50, 100, 200],  # Values are now provided!
    'max_depth': [None, 10, 20, 30],
    'max_features': ['sqrt', 'log2', None],
    'min_samples_split': [2, 5, 10],
    'min_samples_leaf': [1, 2, 4]
}

rf = RandomForestClassifier(random_state=42)
random_search = RandomizedSearchCV(rf, rand_params, n_iter=10, cv=5, n_
jobs=-1, verbose=1, random_state=42)

random_search.fit(X_train, y_train)

print(f"Best parameters: {random_search.best_params_}")

# Evaluate the best model on the test set
best_rf = random_search.best_estimator_
test_accuracy = best_rf.score(X_test, y_test)
print(f"Test Accuracy: {test_accuracy}")
```

Expected Results

```
Fitting 5 folds for each of 10 candidates, totalling 50 fits
Best parameters: {'n_estimators': 100, 'min_samples_split': 5, 'min_
samples_leaf': 2, 'max_features': None, 'max_depth': 30}
Test Accuracy: 0.9
```

Randomized Search can sometimes identify good solutions much faster than Grid Search does by arbitrarily selecting pairs, particularly if only part of the hyperparameter space has ideal values.

Bayesian Optimization

Probabilistic models are employed by Bayesian Optimization to search for the best set of hyperparameters. Rather than randomly or completely exploring the parameter space, it constructs a surrogate model that forecasts which hyperparameters are most promising. Bayesian Optimization concentrates on the regions more likely to produce benefits.

This technique uses an acquisition function. The acquisition function guides the next set of hyperparameters to evaluate based on prior findings. Consequently, Bayesian Optimization is more effective and calls for fewer assessments than either Grid or Randomized Search. The code below uses `Optuna` to tune the hyperparameters using Bayesian Optimization.

Listing 3-4. Hyperparameters: Bayesian Optimization

```
import optuna
from sklearn.ensemble import RandomForestClassifier
from sklearn.model_selection import cross_val_score
from sklearn.datasets import make_classification

# Sample Data (replace with your actual data)
X, y = make_classification(n_samples=100, n_features=10, random_state=42)

def objective(trial):
    n_estimators = trial.suggest_int('n_estimators', 10, 200)
    max_depth = trial.suggest_int('max_depth', 3, 10) #Or suggest_
    categorical(['None', 3, 10]) if you want None as an option
    # ... other hyperparameters ...

    rf = RandomForestClassifier(n_estimators=n_estimators, max_depth=max_
    depth, random_state=42)
    score = cross_val_score(rf, X, y, cv=5, scoring='accuracy').mean()
    # Use cross-validation

    return score

study = optuna.create_study(direction='maximize')  # Maximize accuracy
study.optimize(objective, n_trials=10) # Number of trials

print(f"Best parameters: {study.best_params}")
print(f"Best score: {study.best_value}")
```

Expected Results

```
[I 2025-02-07 20:45:18,226] A new study created in memory with name: no-
name-eb46949e-9fa9-4401-956d-f95efcb95d71
[I 2025-02-07 20:45:18,574] Trial 0 finished with value: 0.93 and
parameters: {'n_estimators': 56, 'max_depth': 4}. Best is trial 0 with
value: 0.93.
[I 2025-02-07 20:45:18,671] Trial 1 finished with value: 0.9299999999999999
and parameters: {'n_estimators': 13, 'max_depth': 9}. Best is trial 0 with
value: 0.93.
[I 2025-02-07 20:45:19,408] Trial 2 finished with value: 0.93 and
parameters: {'n_estimators': 119, 'max_depth': 9}. Best is trial 0 with
value: 0.93.
[I 2025-02-07 20:45:19,975] Trial 3 finished with value: 0.93 and
parameters: {'n_estimators': 93, 'max_depth': 6}. Best is trial 0 with
value: 0.93.
[I 2025-02-07 20:45:20,754] Trial 4 finished with value: 0.9299999999999999
and parameters: {'n_estimators': 120, 'max_depth': 10}. Best is trial 0
with value: 0.93.
[I 2025-02-07 20:45:21,750] Trial 5 finished with value: 0.9400000000000001
and parameters: {'n_estimators': 162, 'max_depth': 5}. Best is trial 5 with
value: 0.9400000000000001.
[I 2025-02-07 20:45:22,869] Trial 6 finished with value: 0.9400000000000001
and parameters: {'n_estimators': 184, 'max_depth': 5}. Best is trial 5 with
value: 0.9400000000000001.
[I 2025-02-07 20:45:23,217] Trial 7 finished with value: 0.93 and
parameters: {'n_estimators': 56, 'max_depth': 5}. Best is trial 5 with
value: 0.9400000000000001.
[I 2025-02-07 20:45:23,314] Trial 8 finished with value: 0.9299999999999999
and parameters: {'n_estimators': 14, 'max_depth': 6}. Best is trial 5 with
value: 0.9400000000000001.
[I 2025-02-07 20:45:23,679] Trial 9 finished with value: 0.93 and
parameters: {'n_estimators': 55, 'max_depth': 8}. Best is trial 5 with
value: 0.9400000000000001.
Best parameters: {'n_estimators': 162, 'max_depth': 5}
Best score: 0.9400000000000001
```

Bayesian Optimization is especially helpful in the case of computationally heavy models because it seeks to reduce the number of evaluations necessary to determine the optimal hyperparameters.

Hyperband

One advanced technique, hyperband, uses early stopping combined with random search to optimize hyperparameters. The primary benefit of hyperband is that it quickly eliminates underperforming hyperparameter sets and apportions more resources to promising ones.

Hyperband evaluates several designs for a restricted number of iterations and then incrementally assigns more iterations to those that do well. By concentrating on the top candidates, this method saves computer resources. The code below shows a use case of hyperband tuning.

Listing 3-5. Hyperparameters: Hyperband

```
import hpbandster.core.nameserver as ns
from hpbandster.core.worker import Worker
from hpbandster.optimizers import HyperBand
import ConfigSpace as CS
import ConfigSpace.hyperparameters as CSH
from sklearn.ensemble import RandomForestClassifier
from sklearn.model_selection import cross_val_score, train_test_split
from sklearn.datasets import make_classification
import logging
import uuid

# Configure logging
logging.basicConfig(level=logging.INFO)

# 1. Configuration Space
config_space = CS.ConfigurationSpace()
n_estimators = CSH.UniformIntegerHyperparameter("n_estimators", lower=10,
upper=200, default_value=100)
max_depth = CSH.UniformIntegerHyperparameter("max_depth", lower=3,
upper=10, default_value=5)
```

```
config_space.add(n_estimators)
config_space.add(max_depth)

# 2. Worker Class (Simplified - No Multiprocessing)
class MyWorker(Worker):  # Inherit from Worker
    def compute(self, config, budget, **kwargs):
        rf = RandomForestClassifier(n_estimators=config["n_estimators"],
                                    max_depth=config["max_depth"],
                                    random_state=42)
        score = cross_val_score(rf, X_train, y_train, cv=3,
        scoring='accuracy').mean()
        return ({'accuracy': score},)  # Must return a tuple

# 3. Sample Data
X, y = make_classification(n_samples=100, n_features=10, random_state=42)
X_train, X_test, y_train, y_test = train_test_split(X, y, test_size=0.2,
random_state=42)

# 4. Run ID
run_id = str(uuid.uuid4())

# 5. Initialize NameServer (Simplified - In-Process)
NS = ns.NameServer(host='127.0.0.1', port=None, run_id=run_id)
NS.start()  # Start the nameserver

# 6. Initialize and Run Worker (Simplified - No separate process)
w = MyWorker(nameserver='127.0.0.1', nameserver_port=NS.port,
id='Worker_0', run_id=run_id)

# 7. Initialize Hyperband
hb = HyperBand(configspace=config_space,
               eta=3,
               min_budget=1,
               max_budget=81,
               nameserver='127.0.0.1',
               nameserver_port=NS.port,
               run_id=run_id)
```

```
# 8. Run Hyperband
hb.run(10)

# 9. Get Best Configuration
id, best_config = hb.get_incumbent()
print(f"Best Configuration: {best_config}")

# 10. Shutdown
hb.shutdown(shutdown_workers=True)
NS.shutdown()

# 11. Evaluate the best model
rf_best = RandomForestClassifier(n_estimators=best_config['n_estimators'],
                                 max_depth=best_config['max_depth'],
                                 random_state=42)
rf_best.fit(X_train, y_train)
test_accuracy = rf_best.score(X_test, y_test)
print(f"Test Accuracy: {test_accuracy}")
```

Expected Results

```
14:28:36 wait_for_workers trying to get the condition
14:28:36 DISPATCHER: started the 'discover_worker' thread
14:28:36 DISPATCHER: started the 'job_runner' thread
14:28:36 DISPATCHER: Pyro daemon running on localhost:16683
14:28:36 DISPATCHER: Starting worker discovery
```

When HpBandSter is employed, these messages are expected and normal. They show the dispatcher is waiting for the worker to connect. The messages signal that the code is operating, but it is waiting for the real hyperparameter optimization to start.

Genetic Algorithms

The genetic algorithm (GA) is designed based on natural selection and evolution. GAs produce a population of possible solutions in hyperparameter tuning and use mutations, crossover, and selection to move toward an ideal collection of hyperparameters. Fitness functions, such as model performance on the validation set, help to assess every generation.[17,18]

Genetic algorithms are efficient in investigating big, complicated hyperparameter spaces and may identify optimal or near-optimal solutions without a thorough search of the full space. Using genetic algorithms is very common in optimization problems, and here it optimizes hyperparameters. There are many similar algorithms, such as simulated annealing, particle swarm optimization, etc., and every year, researchers introduce new algorithms that can be used in hyperparameter optimization.

In this chapter, first, we discussed how to select the most suitable model, and then we optimized the hyperparameters. With these two steps, we have made much progress in model selection and optimization.

Project: Credit Card Fraud Detection and Model Optimization

Fraud detection is very important for ensuring a secure data flow. It is especially more crucial in financial transactions. In the project of this chapter, we practice our learning in a fraud detection problem.

Objective

This project's purpose is to create an end-to-end machine learning pipeline to find counterfeit credit card transactions. By this, you will see how to handle the problem of imbalanced datasets. In this project, you will compare several classification models and refine the chosen model via careful hyperparameter tuning. At the end, you will know the methods and top techniques needed to address real-world classification issues as well as to create, test, and implement an improved forecast model.

Project Description

This project includes using a credit card fraud detection dataset. Naturally, this dataset is imbalanced—fake transactions (the underrepresented class) happen much less than typical ones. We had a comprehensive discussion about the imbalanced dataset in Chapter 2. The project has ten consecutive steps:

There are ten main steps for the project.

1. Create the coding environment by importing essential libraries.
2. **Download the Dataset**: Check for the local availability of the dataset.
3. **Load the File**: Examine and read the dataset.
4. **Manage Imbalanced Data**: Use several methods to balance the classes.
5. **Set a Set of Possible Models**: Define several models.
6. **Make Pipelines for Preprocessing and Modeling**: Standardize features and wrap models.
7. **Perform Model Selection Cross-Validation**: ROC AUC scores allow you to assess models.
8. **Pick the Top-Performing Classifier**: Choose the best model.
9. **Hyperparameter Optimization**: Use `GridSearchCV` and Bayesian Optimization to optimize the parameters of the chosen model.
10. **Save the Optimized Model**: Keep the end model on the disk.

Step 1: Import Necessary Libraries

In this first stage, we import all libraries that are required for our project. These libraries are needed for imbalanced data management (`imblearn`), hyperparameter tuning (`GridSearchCV` and `BayesSearchCV`), visualization (`matplotlib` and `seaborn`), model training (`scikitlearn`), and data manipulation (using `pandas` and `NumPy`). The code below imports the required libraries.

This code downloads all required libraries. If there is no result, it means that everything is fine.

Listing 3-6. Step 1: Import Necessary Libraries

```
import os
import pandas as pd
import numpy as np
```

```
from sklearn.model_selection import train_test_split, GridSearchCV, cross_
val_score
from sklearn.metrics import (
    classification_report, confusion_matrix, roc_auc_score, precision_
    recall_curve,
    auc, f1_score, make_scorer, roc_curve
)
from sklearn.ensemble import RandomForestClassifier, VotingClassifier
from sklearn.linear_model import LogisticRegression
from sklearn.svm import SVC
from sklearn.utils import resample
from imblearn.over_sampling import SMOTE
from imblearn.ensemble import BalancedRandomForestClassifier
from sklearn.pipeline import Pipeline
from sklearn.preprocessing import StandardScaler
from skopt import BayesSearchCV
import matplotlib.pyplot as plt
import warnings
import requests

# Suppress warnings for cleaner output
warnings.filterwarnings('ignore')
```

Expected Results

This code will not produce any results. If there is no output, everything is good.

Step 2: Download the Dataset

This step investigates whether the credit card fraud dataset is present in the local environment. If it is not present, the code downloads the dataset from a publicly available URL. The code below downloads the dataset.

The function `download_dataset()` checks if the file `creditcard.csv` is present locally; if it is not present, it uses the requests library to download the dataset from Google Cloud Storage onto the disk. Informative messages print progress updates to the user. This code is a good practice. In many projects, you may download that data, but after that, if you re-run the code, you do not want to re-download the data.

Listing 3-7. Step 2: Download the Dataset

```
def download_dataset():
    dataset_url = "https://storage.googleapis.com/download.tensorflow.org/
    data/creditcard.csv"
    file_name = "creditcard.csv"

    if not os.path.exists(file_name):
        print(f"Downloading dataset from {dataset_url}...")
        response = requests.get(dataset_url)
        if response.status_code == 200:
            with open(file_name, 'wb') as file:
                file.write(response.content)
            print(f"Dataset downloaded and saved as '{file_name}'.")
        else:
            raise Exception("Failed to download the dataset. Please check
            your internet connection.")
    else:
        print(f"Dataset '{file_name}' already exists. Skipping download.")

print("\nStep 2: Checking for dataset...")
download_dataset()
```

Expected Results

The dataset will either be downloaded, or if it exists, a message will show the state:

```
"Downloading dataset from https://storage.googleapis.com/download.
tensorflow.org/data/creditcard.csv..." followed by "Dataset downloaded and
saved as 'creditcard.csv'."
or
"Dataset 'creditcard.csv' already exists. Skipping download."
```

Step 3: Load the Dataset

After the dataset is downloaded, it will be loaded into a pandas DataFrame. In this step, basic information about the dataset is displayed, including structure and class distribution. The code below loads the dataset.

This step reads the CSV file into a DataFrame using pd.read_csv(). The info() method provides a concise summary of the dataset's dimensions and data types, whereas value_counts() on the 'Class' column shows the disparity between non-fraudulent (0) and fraudulent (1) transactions.

Listing 3-8. Step 3: Load the Dataset

```
print("\nLoading the dataset...")
data = pd.read_csv('creditcard.csv')

# Display basic information about the dataset
print("\nDataset Overview:")
print(data.info())
print("\nClass Distribution:")
print(data['Class'].value_counts())
```

Expected Results

The result consists of

- A concise summary of the dataset's structure (number of rows, columns, and data types)
- A printout of class distribution: the counts of non-fraudulent (0) vs. fraudulent (1) transactions.

```
Loading the dataset...

Dataset Overview:
<class 'pandas.core.frame.DataFrame'>
RangeIndex: 284807 entries, 0 to 284806
Data columns (total 31 columns):
 #   Column  Non-Null Count   Dtype
---  ------  --------------   -----
 0   Time    284807 non-null  float64
 1   V1      284807 non-null  float64
 2   V2      284807 non-null  float64
 3   V3      284807 non-null  float64
 4   V4      284807 non-null  float64
```

```
 5   V5      284807 non-null  float64
 6   V6      284807 non-null  float64
 7   V7      284807 non-null  float64
 8   V8      284807 non-null  float64
 9   V9      284807 non-null  float64
 10  V10     284807 non-null  float64
 11  V11     284807 non-null  float64
 12  V12     284807 non-null  float64
 13  V13     284807 non-null  float64
 14  V14     284807 non-null  float64
 15  V15     284807 non-null  float64
 16  V16     284807 non-null  float64
 17  V17     284807 non-null  float64
 18  V18     284807 non-null  float64
 19  V19     284807 non-null  float64
 20  V20     284807 non-null  float64
 21  V21     284807 non-null  float64
 22  V22     284807 non-null  float64
 23  V23     284807 non-null  float64
 24  V24     284807 non-null  float64
 25  V25     284807 non-null  float64
 26  V26     284807 non-null  float64
 27  V27     284807 non-null  float64
 28  V28     284807 non-null  float64
 29  Amount  284807 non-null  float64
 30  Class   284807 non-null  int64
dtypes: float64(30), int64(1)
memory usage: 67.4 MB
None

Class Distribution:
Class
0    284315
1       492
Name: count, dtype: int64
```

Step 4: Handle Imbalanced Data Using Multiple Techniques

Since fraudulent transactions are a rare phenomenon, it becomes imperative to rebalance the dataset. In this step, SMOTE for synthetic oversampling, random undersampling for the majority class, and preparation of class weights for models are applied. We will use different techniques to cover more topics that we discussed earlier Chapter 2. The code below handles imbalanced data.

Since the dataset is intensely unbalanced, it's important that we use some techniques to help the model learn fairly from both classes. First, the dataset is split into training and testing sets in a way to preserve the original class distribution.

Listing 3-9. Step 4: Handle Imbalanced Data

```
print("\nHandling imbalanced data using multiple techniques...")

# Split the data into training and testing sets
X = data.drop('Class', axis=1)
y = data['Class']
X_train, X_test, y_train, y_test = train_test_split(X, y, test_size=0.3,
random_state=42, stratify=y)

# Technique 1: SMOTE (Synthetic Minority Over-sampling)
smote = SMOTE(random_state=42)
X_train_smote, y_train_smote = smote.fit_resample(X_train, y_train)

# Technique 2: Undersampling (Random Undersampling)
X_train_undersampled, y_train_undersampled = resample(
    X_train[y_train == 0], y_train[y_train == 0],
    replace=False, n_samples=len(y_train[y_train == 1]), random_state=42
)
X_train_undersampled = pd.concat([X_train_undersampled, X_train[y_
train == 1]])
y_train_undersampled = pd.concat([y_train_undersampled, y_train[y_
train == 1]])
```

```
# Technique 3: Class Weighting (for models that support it)
class_weights = {0: 1, 1: len(y_train[y_train == 0]) / len(y_train[y_
train == 1])}
```

Expected Results

```
Handling imbalanced data using multiple techniques...
```

The output expected from this step is that several balanced forms of training data (over- and undersampled) should be generated. In addition, class weighting pieces of information should be prepared. This step does not have a visible output; just a few debug messages may appear.

Step 5: Define Multiple Models for Comparison

To determine the best response for detecting fraud, multiple candidate models will be defined so we can have a robust comparison by cross-validation. The code below defines the models.

In this step, we define a dictionary of candidate models that will be compared using cross-validation. The selected models include logistic regression, a standard Random Forest, a Balanced Random Forest (specifically meant for handling imbalanced data), and a support vector machine with probability estimates enabled. Each model is instantiated with a fixed random state, ensuring reproducibility of the results. This collection allows for comparison between traditional models and those that were explicitly designed for imbalanced datasets.

Listing 3-10. Step 5: Define Multiple Models for Comparison

```
print("\nDefining models for comparison...")
models = {
    "Logistic Regression": LogisticRegression(max_iter=1000, random_
    state=42),
    "Random Forest": RandomForestClassifier(random_state=42),
    "Balanced Random Forest": BalancedRandomForestClassifier(random_
    state=42),
    "Support Vector Machine": SVC(probability=True, random_state=42)
}
```

Expected Results

```
Defining models for comparison...
```

After executing this step, the variable models will contain four candidate classifiers. If there is no printed output, it means everything has been set correctly. You can define more models in your code.

Step 6: Create Pipelines for Preprocessing and Modeling

In this step, pipelines are created that integrate feature scaling with each of the candidate models. This guarantees a standard input format for model evaluations. The code below created the pipelines.

For each candidate model, a pipeline is created that first implements standard scaling for feature normalization before fitting the classifier. Because of this, all models receive data in a consistent, scaled format, with which comparisons can be made in a completely fair way.

Listing 3-11. Step 6: Create Pipelines for Preprocessing and Modeling

```
print("\nCreating pipelines for preprocessing and modeling...")
pipelines = {}
for model_name, model in models.items():
    pipelines[model_name] = Pipeline([
        ('scaler', StandardScaler()),  # Feature scaling
        ('model', model)
    ])
```

Expected Results

```
Creating pipelines for preprocessing and modeling...
```

The result of this step is a dictionary of pipelines, where each key (model name) maps to a pipeline that includes both preprocessing (scaling) and a machine learning classifier. There is no visible output on this stage; the pipelines are computed for model evaluation now.

Step 7: Perform Cross-Validation for Model Selection

In this step, each pipeline is evaluated using cross-validation in conjunction with ROC AUC as a performance score. This score is especially suited for imbalanced classification problems. The code below performs cross-validation.

Each pipeline is cross-validated over five folds, with ROC AUC as the scoring metric. The previously defined class weights are used for logistic regression. The mean ROC AUC scores are stored in a dictionary (cv_results) and printed to compare their performance directly. Read the above code, see how Python libraries make machine learning implementation easy.

Listing 3-12. Step 7: Perform Cross-Validation for Model Selection

```
print("\nPerforming cross-validation for model selection...")
cv_results = {}
for model_name, pipeline in tqdm(pipelines.items(), desc="Cross-validating
models", unit="model"):
    if model_name == "Logistic Regression":
        pipeline.set_params(model__class_weight=class_weights)  # Apply
        class weighting

    # Perform cross-validation with progress bar for each fold
    scores = cross_val_score(pipeline, X_train, y_train, cv=5,
    scoring='roc_auc')
    cv_results[model_name] = scores.mean()
    print(f"\n{model_name} - Mean ROC AUC: {scores.mean():.4f}")
```

Expected Results

```
Performing cross-validation for model selection...
Cross-validating models:  25%|██████████
            | 1/4 [00:03<00:09,  3.21s/model]

Logistic Regression - Mean ROC AUC: 0.9821
Cross-validating
models:  50%|████████████████████
██                                    | 2/4 [10:44<12:37, 378.79s/model]
```

```
Random Forest - Mean ROC AUC: 0.9550
Cross-validating models:  75%|████████████████████████████████████████|
3/4 [11:03<03:34, 214.30s/model]

Balanced Random Forest - Mean ROC AUC: 0.9810
Cross-validating models: 100%|████████████████████████████████████████████████|
4/4 [1:18:55<00:00, 1183.92s/model]

Support Vector Machine - Mean ROC AUC: 0.9521
```

The mean ROC AUC for each model will be visible on the console, along with a progress bar.

Step 8: Select the Best Model Based on Cross-Validation Results

The model with the highest average ROC AUC score is selected as the best candidate for further optimization. Do you know why we did not consider accuracy as a measure to select the best model? The code below selects the best model based on cross-validation.

Using the dictionary of cross-validation results, the highest mean ROC AUC is found from the selected model. The pipeline related to this model is stored as `best_pipeline`, and its name is printed. This step ensures that hyperparameter tuning will target the most promising candidate.

Listing 3-13. Step 8: Select the Best Model Based on Cross-Validation Results

```
best_model_name = max(cv_results, key=cv_results.get)
best_pipeline = pipelines[best_model_name]
print(f"\nBest Model Selected: {best_model_name}")
print("\nPerforming hyperparameter optimization for the best model...")
```

Expected Results

```
Best Model Selected: Logistic Regression
```

The output shows the selected model as the best performer.

Step 9: Hyperparameter Optimization Using GridSearchCV and Bayesian Optimization

In this step, we optimize the hyperparameters using Grid Search and Bayesian Optimization.

Depending on the selected model, a parameter grid is defined. For instance, if the best model is Random Forest, the grid may include various options for number of trees, maximum depths, and minimum samples required to split a node (in our example, if the best model were "Balanced Random Forest," the structure of the current code would pass to the SVM one by default). This could be improved by adding another condition checking for Balanced Random Forest to select the appropriate hyperparameter sets for tuning. A combination of two hyperparameter optimization strategies is undertaken:

1. **GridSearchCV**: It evaluates each combination of the parameters grid using exhaustive cross-validation.
2. **BayesSearchCV**: Performs an efficient exploration of the hyperparameter space by exploiting Bayesian Optimization.

With these tuning methods, the oversampled set (X_train_smote, y_train_smote) is used, so that the minority class is well represented during tuning. The best hyperparameters from both methods are used at this stage to select the final model based on which method scored higher.

Listing 3-14. Step 9: Hyperparameter Optimization Using GridSearchCV and Bayesian Optimization

```
print("\nPerforming hyperparameter optimization for the best model...")

if best_model_name == "Random Forest":
    param_grid = {
        'model__n_estimators': [50, 100, 200],
        'model__max_depth': [None, 10, 20],
        'model__min_samples_split': [2, 5, 10]
    }
elif best_model_name == "Logistic Regression":
    param_grid = {
        'model__C': [0.01, 0.1, 1, 10],
```

```
        'model__solver': ['liblinear', 'lbfgs']
    }
else:  # Support Vector Machine or other models
    param_grid = {
        'model__C': [0.1, 1, 10],
        'model__kernel': ['linear', 'rbf']
    }

# GridSearchCV
grid_search = GridSearchCV(best_pipeline, param_grid, cv=5, scoring='roc_
auc', n_jobs=-1)
grid_search.fit(X_train_smote, y_train_smote)

# Bayesian Optimization (BayesSearchCV)
bayes_search = BayesSearchCV(best_pipeline, param_grid, n_iter=10, cv=5,
scoring='roc_auc', n_jobs=-1)
bayes_search.fit(X_train_smote, y_train_smote)

print(f"Best Hyperparameters (GridSearch): {grid_search.best_params_}")
print(f"Best Hyperparameters (Bayesian): {bayes_search.best_params_}")

# Choose the best estimator from both methods
optimized_model = grid_search.best_estimator_ if grid_search.best_score_ >
bayes_search.best_score_ else bayes_search.best_estimator_
```

Expected Results

The best hyperparameters obtained using both GridSearchCV and Bayesian Optimization will be printed in the console. Figure 3-1 and Figure 3-2 show a confusion matrix and ROC curves, respectively.

The final optimized model (named `optimized_model`) will be available for further evaluation.

```
Evaluating the optimized model on the test set...

Confusion Matrix:
[[64168 21127]
 [    7   141]]
```

```
Classification Report:
              precision    recall  f1-score   support

           0       1.00      0.75      0.86     85295
           1       0.01      0.95      0.01       148

    accuracy                           0.75     85443
   macro avg       0.50      0.85      0.44     85443
weighted avg       1.00      0.75      0.86     85443

ROC AUC Score:
0.9645609514197941
Precision-Recall AUC: 0.7319
```

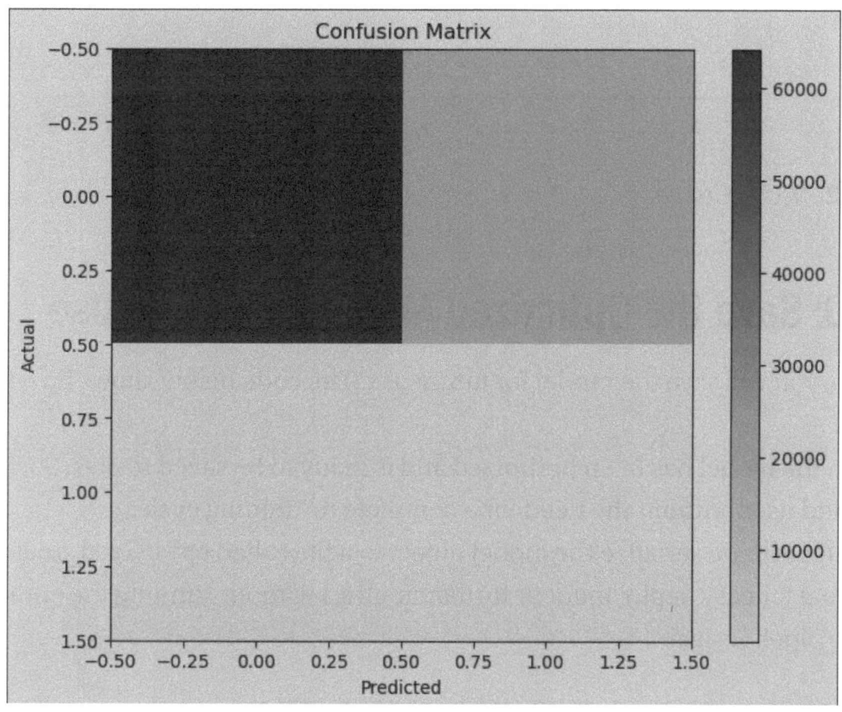

Figure 3-1. Confusion Matrix

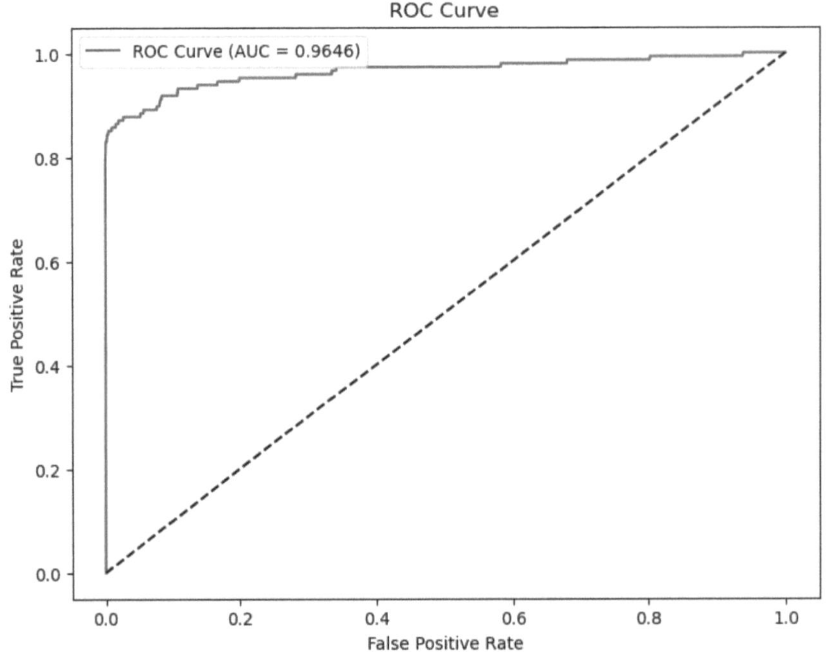

Figure 3-2. *ROC Curve*

Step 10: Save the Optimized Model for Future Use

In this step, you will save the model for future use. The code below shows how to save the model.

Finally, the model has been optimized and is ready to be saved to disk. So, it can be retrieved and used without the need for a complete re-training cycle.

Using `joblib`, we serialize the model object to a file called `optimized_model.pkl`, thus allowing for easy deployment or further analysis without running the entire training and tuning pipeline again.

Listing 3-15. Step 10: Save the Optimized Model for Future Use

```
import joblib
print("\nSaving the optimized model...")
joblib.dump(optimized_model, 'optimized_model.pkl')
print("Model saved as 'optimized_model.pkl'")
```

Expected Results

```
Saving the optimized model...
Model saved as 'optimized_model.pkl'
```

A message saying that the optimized model is stored will be printed:

Conclusion

In the project of this chapter, we developed a complete machine learning pipeline to solve a challenging fraud detection problem. We started by downloading and loading the credit card transactions dataset into the project, and we handled imbalanced data. Afterward, we defined and compared several candidate models using robust cross-validation metrics. The best performing model was optimized through both Grid Search and Bayesian Optimization, and its performance was evaluated using multiple metrics; also, some visualizations were generated. Finally, we saved the optimized model on disk for future use. In many projects, especially research ones, you may be asked to improve a current model. For this case, you should consider the current model metrics as a baseline, and then whenever you change something, such as hyperparameters, you can compare the performance with your defined baseline to see whether it has improved or not.

Summary

In this chapter, we primarily introduced different types of models, and we talked about various methods for selecting the most suitable model for a problem. In some code examples, we automated the model selection.

Afterward, we talked about hyperparameters, their effect, and how to tune them. We discussed various techniques of hyperparameter optimization and their advantages and disadvantages. Also, we talked about some procedures that help to select the most suitable model and its hyperparameters for a specific project.

With these implementations, you will be able to build a machine learning model that would presumably perform well across all the classes, incorporating the underrepresented minority class. So, until now, first in Chapter 2, we cleaned the data and prepared that for training the model; in this chapter, we find out how to select the best model and hyperparameters. Then, in the next chapter, we will learn how to design a scalable and modular ML system.

CHAPTER 4

Building Scalable and Modular ML Systems

One of the fundamental ideas for creating scalable and maintainable machine learning (ML) systems is modularization. Modularity separates a system into smaller, independent components, making it easier to scale, debug, update, and manage. Modular design promotes improved group cooperation, component reuse, and adaptability to new features or changes. In this chapter, we will discuss the main ideas of modular design, how to apply modular structures in ML systems, and their advantages in scaling and supporting intricate solutions. If you are working on a real-world project, modularity is essential; without it, integration and maintenance challenges will grow substantially over time. The complexity of integrating new features, updating components, and reusing code across projects can be dealt with by means of modular design principles. Creating ML systems that can evolve without compromising performance, maintainability, or clarity depends on these fundamental ideas.

Highlighting the need for modularity and microservices architecture, in this chapter, we cover the design and deployment of scalable and modular machine learning (ML) systems. We start with setting up fundamental pieces of modular design and considering their use in machine learning systems to improve scalability and maintainability. Next, we introduce microservices, explaining how various components fit into a microservices-based ML ecosystem and outlining their benefits such as enhanced flexibility, isolation, and scalability. We then cover techniques for handling large datasets with microservices, leveraging GPUs for accelerated training and inference, and incorporating design patterns that support robust microservice-driven ML architectures. Finally, we address the challenges associated with adopting microservices in ML and examine how cloud platforms support the scaling of these distributed systems. Together, these practices provide practical guidance for building efficient, scalable, and modular ML systems that can evolve while minimizing interdependencies.

M. R. Mahdiani, *Mastering Machine Learning Architecture and Solutions*,
https://doi.org/10.1007/979-8-8688-2527-9_4

Modular Design Principles

There are some fundamental concepts that we need to consider while designing a modular system. In this section, we focus on these concepts, beginning with the separation of concerns.

Separation of Concerns

Separation of concerns (SoC) is a design principle that breaks a complex problem into smaller, more manageable parts. Each module or component of the system should have one responsibility. Data preprocessing, model training, evaluation, and deployment should be seen as separate components. This separation lets every module be independently developed, tested, and upgraded.

For instance, a project can have three modules, including data processing, model training, and a deployment module.

The data preprocessing module manages all feature extraction, data cleaning, and transformation, while the model training module does the verification, hyperparameter tuning, and model training control. The deployment module manages the deployment of a trained model in a real-world setting.

Dividing these modules will help to simplify maintenance by ensuring that edits made to one part (feature extraction, for example) do not affect others. The code below shows a sample that violates the separation of concerns. After fixing the code, you can see that it has been improved and follows the separation of concerns.

Listing 4-1. Sample Code That Violates the Separation of Concerns

```
import pandas as pd
from sklearn.linear_model import LogisticRegression
from sklearn.metrics import accuracy_score

# Everything mixed together
data = pd.read_csv("data.csv")
X = data.drop("target", axis=1).fillna(0)
y = data["target"]
X = (X - X.min()) / (X.max() - X.min())
model = LogisticRegression()
```

```
model.fit(X, y)
preds = model.predict(X)
print(f"Accuracy: {accuracy_score(y, preds)}")
```

The above code violates the separation of concerns; as the project grows, maintaining and scaling code similar to the above is very difficult. The code below shows a clean sample code that follows the separation of concerns principle.

Listing 4-2. Sample Code That Follows the Separation of Concerns

```
import pandas as pd
from sklearn.linear_model import LogisticRegression
from sklearn.metrics import accuracy_score

# Preprocessing module
def preprocess_data(data):
    X = data.drop("target", axis=1).fillna(0)
    return (X - X.min()) / (X.max() - X.min()), data["target"]

# Training module
def train_model(X, y):
    model = LogisticRegression()
    model.fit(X, y)
    return model

# Evaluation module
def evaluate_model(model, X, y):
    preds = model.predict(X)
    return accuracy_score(y, preds)

# Usage
data = pd.read_csv("data.csv")
X, y = preprocess_data(data)
model = train_model(X, y)
accuracy = evaluate_model(model, X, y)
print(f"Accuracy: {accuracy}")
```

High Cohesion and Low Coupling

Modular systems have two critical properties: high cohesion and low coupling. High cohesion means that the elements within a module are closely related and focused on a single, well-defined purpose. Low coupling refers to the level of dependency between modules; ideally, modules should operate as independently as possible.

In an ML system, high cohesion ensures that a data preprocessing module handles only tasks related to data cleaning, transformation, and feature preparation. Low coupling means the preprocessing and model training modules interact through well-defined interfaces, allowing either module to be modified or replaced without affecting the other.A model training module should get preprocessed data from the data preprocessing module rather than access raw data straight from a database. This establishes a definite boundary and lessens interdependence. Like before, below, we show you two sample codes, one where the principle of high cohesion and low coupling is not considered, and after that, you see the improved code. Try to explain to yourself why the corrected code follows the principle while the original code does not.

Listing 4-3. Sample Code Without Considering High Cohesion and Low Coupling

```
import pandas as pd
from sklearn.linear_model import LogisticRegression

# One function does everything and accesses data directly
def process_and_train():
    data = pd.read_csv("data.csv")  # Direct data access
    X = data.drop("target", axis=1)
    X["feature1"] = X["feature1"].fillna(0)  # Specific preprocessing
    y = data["target"]
    model = LogisticRegression()
    model.fit(X, y)
    return model

model = process_and_train()
```

Listing 4-4. Sample Code After Improving It to Follow High Cohesion and Low Coupling

```
import pandas as pd
from sklearn.linear_model import LogisticRegression
# Cohesive preprocessing module

def preprocess(X):
    return X.fillna(0)

# Cohesive training module
def train_model(X, y):
    model = LogisticRegression()
    model.fit(X, y)
    return model

# Usage with loose coupling
data = pd.read_csv("data.csv")
X_raw = data.drop("target", axis=1)
y = data["target"]
X_processed = preprocess(X_raw)
```

model = train_model(X_processed, y) Reusability

Modular design offers great reusability. Modules created to accomplish purposes can be reused among various projects or situations. For example, a data validation module that searches for outliers and missing values could be applied in several machine learning pipelines without any change.

Development time is saved with reusable components, and the chances of errors being introduced are lowered. Moreover, as standardized modules can be shared and employed in various ML tasks, they also help preserve consistency across projects. For instance, regardless of the kind of model being used, a feature scaling module that normalizes features to a common scale can be reused in several ML projects. Again, we show two sample codes. The first one does not follow the reusability principle, and the second one does.

Listing 4-5. Sample Code Without Considering Reusability

```
import pandas as pd
from sklearn.preprocessing import MinMaxScaler

# Hardcoded, project-specific preprocessing
data = pd.read_csv("data.csv")
X = data.drop("target", axis=1)
scaler = MinMaxScaler()
X_scaled = scaler.fit_transform(X)  # Tied to this dataset
print(X_scaled)
```

Listing 4-6. Sample Code That Considers Reusability

```
import pandas as pd
from sklearn.preprocessing import MinMaxScaler

# Reusable preprocessing function
def scale_features(X, scaler=None):
    if scaler is None:
        scaler = MinMaxScaler()
        scaler.fit(X)
    return scaler.transform(X), scaler

# Usage
data = pd.read_csv("data.csv")
X = data.drop("target", axis=1)
X_scaled, scaler = scale_features(X)
print(X_scaled)
# Reusable scaler for new data
new_data = pd.read_csv("new_data.csv")
new_scaled = scale_features(new_data, scaler)[0]
```

Abstraction

Modular design allows for abstraction, which exposes only the essential features of a module while concealing its internal implementation details. Developers may interact with each module via clearly defined interfaces, thanks to this abstraction, free of concern about the internal operations. An example of abstraction is a model training module that turned into a function that accepts data and hyperparameters and produces a trained model. Users are kept unaware of the internal processes, such as batch training, optimizer selection, and convergence criteria, that define the system. The following shows an example that does not follow abstraction, then the next code shows how it can be fixed.

Listing 4-7. Sample Code Without Considering Abstraction

```
import pandas as pd
from sklearn.linear_model import LogisticRegression

# Exposed implementation details
data = pd.read_csv("data.csv")
X = data.drop("target", axis=1).fillna(0)
y = data["target"]
model = LogisticRegression(max_iter=100, solver='lbfgs')
model.fit(X, y)
print(model.coef_)
```

Listing 4-8. Improved Sample Code for Considering Abstraction

```
import pandas as pd
from sklearn.linear_model import LogisticRegression

# Abstracted training function
def train_model(X, y, params={'max_iter': 100}):
    model = LogisticRegression(**params)
    model.fit(X, y)
    return model

# Usage
data = pd.read_csv("data.csv")
X = data.drop("target", axis=1).fillna(0)
y = data["target"]
model = train_model(X, y)
print(model.coef_)
```

Scalability

An important aspect of designing modular ML systems is scalability. The good thing about modular ML systems is that they are separate parts. Each part of modular ML systems can be scaled independently. For example, if the bottleneck is in data preprocessing (due to an increase in data volume or more users), the only part that needs to be scaled is the preprocessing module. For this reason, I suggest using a modular system in ML (and also non-ML) projects.

The data ingestion module can be scaled by load distribution across several nodes. In this case, the model training module will remain unaffected. This type of scaling ensures efficient use of computational resources and minimizes operational costs. In future chapters, we will discuss more about operational cost management. There are several tools for parallel data processing. One of them is `PySpark`. In Chapter 2, we got familiar with `PySpark`. Here, we show two sample codes. One that is not using `PySpark` and is not scalable and the second one that we use `PySpark` to make scalable.

Listing 4-9. Non-scalable Code

```
import pandas as pd

# Non-scalable preprocessing
data = pd.read_csv("data.csv")
X = data.drop("target", axis=1)
for col in X.columns:
    X[col] = X[col].fillna(X[col].mean())  # Loops over columns in memory
print(X)
```

Listing 4-10. Scalable Code using pyspark

```
from pyspark.sql import SparkSession
from pyspark.sql.functions import col, mean

# Scalable preprocessing with PySpark
spark = SparkSession.builder.appName("ScalableML").getOrCreate()
data = spark.read.csv("data.csv", header=True, inferSchema=True)
X = data.drop("target")
means = {col: X.select(mean(col).alias("mean")).collect()[0]["mean"] for
col in X.columns}
```

```
X = X.na.fill(means)
X.show()
spark.stop()
```

Implementing Modular Design in ML Systems

For implementing modular design principles in ML systems, you can use some frameworks and tools that support modular architecture. In the following, some practical approaches are listed:[2]

Pipeline Frameworks: Tools such as Apache Kubeflow, Airflow, and Luigi help in creating reusable and modular projects. Using these tools, each part of the pipeline can be developed as an independent module; these frameworks manage orchestration, scheduling, dependencies, and workflow execution in a structured manner.

Containerization: Containers, such as Docker, help to package each module. Packaging allows easy deployment and scaling without worrying about dependencies. Containers provide an isolated environment for each module on the same or different computers. These techniques ensure that dependencies will not interfere with each other.

Microservices Architecture: Microservices architecture is my favorite for big projects. In microservices architecture, each module (which can be containerized for easier deployment) is deployed as an independent service that communicates with other services via APIs. This approach leads to low coupling and a highly scalable system. In this way, each service can be updated or scaled independently without affecting the whole system. Each service can be updated or scaled independently. When many services exist, orchestration tools such as Kubernetes are often needed for deployment, scaling, and monitoring.
Example: As mentioned earlier, in an ML system developed using microservices, the data preprocessing module can be hosted as a separate service. This service can be accessed via an API endpoint. Also, the model inference module can be another

service that takes input data and returns predictions. Table 4-1 shows more details and compares the tools and frameworks that can be used in modular design.

***Table 4-1.** Techniques to Implementing Modular Design in ML Systems*

Technique	Description	Tools and Technologies	Advantages	Disadvantage
Pipeline Frameworks	Create reusable and modular workflows for orchestrating ML tasks	Apache Airflow, Kubeflow, Luigi	Reusable components, easy orchestration	Would be complex for big systems
Containerization	Packages the modules into isolated environments for scalability and easy deployment	Docker	Easy deployment, isolated environments	There is a need for container management tools
Microservices Architecture	Deploy each module as an independent service for more flexibility and low coupling	REST APIs, Kubernetes	Flexible module updates, independent scalability	Increases the operational complexity
Serverless Computing	Allow automatic scaling, modular components as serverless functions	AWS Lambda, Azure Functions	Reduced infrastructure management, automatic scaling	Stateless, limited execution time
Modular Libraries	Utilize reusable libraries for common ML tasks, for example, preprocessing and model evaluation	scikit-learn, TensorFlow	Well-tested, reusable across projects	Less customization in some cases
Cloud-Based Orchestration	Use cloud services for building modular workflows	AWS Step Functions, GCP Composer	Built-in monitoring, scalability	High costs, vendor lock-in
Function-as-a-Service (FaaS)	Modularize ML tasks into independent functions, callable as needed	AWS Lambda, Google Cloud Functions	Highly scalable, cost-effective	Limited execution time, cold start latency

Benefits of Modular Design

In this section, we briefly discuss how modular design can benefit projects, especially ML ones:

Easy Maintenance: Maintaining modular systems is easier. It is because each module is independent from the other modules and is developed for a specific task. In modular design, whenever bugs or issues arise, that specific module is isolated. Thus, the problem can be fixed without impacting the other modules or the entire system.

Team Collaboration: Modular design develops better team collaboration. It is because various teams can work on different modules without interfering with or affecting each other's work. For instance, the data engineering team can work on data preprocessing while the machine learning team works on model training. In this way, as the project is broken into smaller pieces, managing is easier.

Scalability: With a modular design, each component can be scaled separately as needed. This allows for more efficient resources and cost management. In this way, modules that require more computation would be scaled independently.

Flexibility and Reusability: Modules can be reused across different projects or in a single project. This leads to more consistency, saving time, and avoiding duplication. In addition, modular systems are more flexible. Because each component can be easily changed, replaced, or upgraded without the need to rewrite or test the entire system.

Microservices in ML

Microservices architecture has become a common design pattern in modern software engineering, especially big projects. Microservice is commonly used in machine learning (ML) systems. Microservices in ML involve breaking down a monolithic ML pipeline into several independent deployable services that each handle a specific task. These tasks include data preprocessing, model training, and inference (running the model and getting results). This architecture helps in making ML systems more scalable, maintainable, and more suitable for rapid updates. Another benefit of microservice is that it enables teams to use various technologies for different components, which provides more flexibility for more efficient tool selection.[17]

In this section, we discuss the fundamental concepts, benefits, and challenges of microservices in ML. Additionally, we will cover the methods for handling big data, the role of GPUs, and the importance of cloud services in scaling microservices-based ML systems.

What Are Microservices in ML?

Earlier, we discussed the general use of microservice. We stated that microservices architecture is the practice of breaking down an application into a collection of loosely coupled services. In the context of ML, each part of the ML pipeline—including data ingestion, feature extraction, model training, and prediction—can be developed, deployed, and scaled independently in various services. Note that each service is self-contained, which means it contains its own dependencies, runtime, and environment.

Some Examples of Microservices in ML

1. **Data Preprocessing Service**: This service handles tasks related to initial data preprocessing such as data cleaning, feature engineering, and data transformation.
2. **Training Service**: Receives preprocessed data and use it to train the model.
3. **Inference Service**: Gets new input data and runs the trained model to produce results.
4. **Monitoring Service**: Checks the performance and health of the model in production.

Benefits of Microservices in ML

We discussed the general benefits of microservices; here, we focus on their advantages specifically in ML:

Scalability: As each service can be scaled independently, components that need more computation (such as model training) can utilize more resources without affecting other services. For example, we can scale model training while giving less resources to inference service.

Technology Flexibility: Each service can be developed using the best-suited technology or programming language for that service. For instance, data preprocessing can be done in Python, leveraging its rich data analysis libraries, while inference can use C++ for faster execution.[20,21]

Ease of Deployment and Maintenance: Independent services make deployments less risky and reduce the danger of system crashes. Because updates to one service do not impact other services. Additionally, modularity helps with faster bug finding and issue resolving.

Handling Big Data with Microservices

Efficient big data handling is one of the main challenges in building scalable ML systems. Microservices help by distributing the processing workload of big data across several services. This results in more resilient and efficient systems. In the next part, we talk more about efficient data handling.

Techniques for Handling Big Data

In ML systems, we often deal with big data, and managing it is one of the main challenges. Here, we discuss various methods for big data handling.

Data Partitioning: In this method, data is partitioned into several portions. This data can be processed in parallel using several data preprocessing services, in which each service handles a part of the dataset.

Streaming Data Processing: By integrating tools like Apache Kafka into the microservices architecture, real-time data can flow into preprocessing or inference services. This allows ML systems to process incoming data continuously without waiting for batch jobs.

Distributed Storage: Cloud-based storage solutions including **Google Cloud Storage and AWS S3** can store and manage big datasets. Each service can access the distributed storage without the need for local copies of the entire dataset.

Batch Processing: Large datasets can also be processed in smaller, manageable batches using frameworks like Dask or Apache Spark. Batch processing allows for efficient use of computational resources and easier error handling.

Table 4-2 shows techniques and tools for developing microservices in machine learning.

Table 4-2. *Approaches for Handling Big Data in Microservices-Based ML Systems*

Approach	Description	Tools and Technologies
Data Partitioning	Dividing large datasets for parallel processing	Apache Hadoop, Spark, Kubernetes
Streaming Data	Real-time data processing	Apache Kafka, Apache Flink
Distributed Storage	Cloud-based storage for easy data access	Google Cloud Storage, HDFS, AWS S3
Batch Processing	Processing large datasets in smaller batches	Dask, Apache Spark

Utilizing GPUs for Training and Inference

Training ML models, such as deep learning models, is computationally expensive. Using **Graphics Processing Units (GPUs)** is a well-known approach to speed up model training. In a microservices-based ML system, GPUs can be used in specific services that require heavy computation.[22]

For instance, the **training service** can be deployed on nodes with attached GPUs, to benefit the computationally intensive training process from parallelism. GPU-powered microservices can be managed using orchestration tools like **Kubernetes**, which handle resource allocation and scheduling across multiple services. Efficiently.

Design Patterns for Microservices in ML

Microservices-based ML systems can use various design patterns to address some common challenges. These challenges include, but are not limited to, data consistency, fault tolerance, and inter-service communication. In this section, we will discuss some design patterns that can be used in microservices ML systems:

Saga Pattern: The saga pattern is used for keeping data consistency over several services, which is particularly useful in long transactions. This pattern is helpful in ML workflows that involve multiple microservices. For scenarios in which data must be processed and updated in sequence, such as data preprocessing, training, and model deployment, the sage pattern can be useful.

Circuit Breaker Pattern: The circuit breaker pattern is used for preventing the failure of a network or system from cascading across multiple services. For instance, if the model training service gets overloaded or unavailable, or crashes, the circuit breaker stops other services, and they will not send requests. This gives enough time to the system to recover.

API Gateway Pattern: An API gateway is a single-entry point for clients to talk with a microservice. In an ML system, an API gateway routes client requests to the appropriate desired services such as data preprocessing, model training, and inference. This design simplifies the client interactions and provides a unified interface. Many ML services, including LLMs, use API gateways.

Event-Driven Architecture: In event-driven design, pattern services can communicate asynchronously via events. For example, when the data preprocessing service completes a task, the service will trigger an event that will notify the training service to start the model training. This approach improves system responsiveness and decouples the services.

Sidecar Pattern: The sidecar pattern is used for increasing the capabilities of microservices while keeping the main service logic intact. For ML services, a sidecar can handle tasks such as monitoring, logging, or metrics collection while leaving the main service (e.g., inference or training) to concentrate solely on the core functionality.

Role of Cloud Services in Scaling Microservices-Based ML Systems

Cloud services have an important role in scaling and deploying microservices-based ML systems. By using cloud infrastructure, companies can dynamically allocate the compute, storage, and networking resources required at different stages of the ML pipeline. This approach reduces both downtime and operational costs. There are various cloud systems. Here, we just talk about a few of them that are more popular in ML systems:

Compute Services: Cloud platforms like **Google Compute Engine, AWS EC2**, and **Azure Virtual Machines** offer on-demand allocation of compute resources. For instance, training services can use instances that are optimized for GPU usage, while inference services can use a combination of GPU and CPU-optimized instances which are more cost-efficient. For very large models, inference itself can also require GPU acceleration.

Serverless Services: Serverless options like **Google Cloud Functions or AWS Lambda** allow some ML tasks to run as short-lived, independent services. This approach is useful for inference, where the system can return results instantly upon receiving new data.

Orchestration Tools: Tools like **Kubernetes** help in orchestrating and managing multiple microservices. Kubernetes can manage containerized services efficiently, ensuring high availability and automated scaling as needed.

Be careful about using cloud services; these services can go wild in pricing. Misconfigured autoscaling or unused high-end GPU instances can significantly increase bills. Chapter 5 will cover cloud cost management strategies in detail. Table 4-3 *outlines the features and use cases of different cloud platforms that can* be used in microservices-based ML architecture.

Table 4-3. *Comparison of Cloud Services for Microservices-Based ML Systems*

Cloud Service	Description	Tools Available	Use Case
AWS	Offers a wide range of ML services and tools	EC2, S3, Lambda, SageMaker	Compute, storage, serverless services, ML model training
Google Cloud Platform	Provides AI-specific services with scalability	Compute Engine, GCS, Vertex AI	Scalable compute, big data processing, ML model training
Microsoft Azure	Integrated services with enterprise support	Azure ML, Virtual Machines, Blob Storage	Orchestration, enterprise-level ML deployments, and scaling
IBM Cloud	Focuses on AI-driven enterprise solutions	Watson Machine Learning	AI-powered microservices, model training and inference
Alibaba Cloud	Offers cloud computing and data storage	PAI, ECS, OSS	Cloud storage, compute resources, and ML pipeline management

Challenges in Adopting Microservices for ML

While the microservices approach offers considerable benefits, it comes with its own set of challenges.[23] In this section, we discuss the most common challenges of microservices, and we offer some solutions.

Increased Complexity: Breaking down the ML system into microservices increases the complexity of managing inter-service communication, data consistency, and system monitoring. Assume that instead of one service, several services are running, and each of them needs separate maintenance and monitoring. If you update the code of one service, you should check its effect on other services and change them accordingly. So, if the project is not big enough, using microservices is not recommended.

Latency Issues: As microservices communicate over a network, latency issues are a common problem. Latency gets worse if services are spread across multiple geographical regions. API calls also introduce limits related to response time and data transfer rates. These delays can affect real-time ML systems, especially those requiring fast inference.

Model Versioning: Deploying and managing different versions of ML models across multiple microservices can be difficult. To overcome this issue, a proper version control system must be in place. Without good versioning, inconsistency and confusion over various versions would create serious problems.

Managing Dependencies

Effective management of dependencies is important in designing modular and scalable ML systems. This dependency management is more crucial in microservices architecture. Each microservice in an ML system has its own dependencies. These dependencies could range from Python packages and data preprocessing libraries to deep learning frameworks used for model training or inference. Proper dependency management ensures reproducibility, consistency, and reliable deployment across environments. Without it, systems can easily break. I've seen developers report that code which worked perfectly a few months earlier suddenly fails; in many cases, this happens because package versions changed or dependencies were not pinned or isolated correctly.

Now you know the importance of dependency management. So, we proceed with more details for effectively managing the dependencies.

Containerization with Docker

Docker is one of the most common tools for managing dependencies in microservices-based ML systems. With Docker, each microservice can be packed into a container that includes the code, libraries, runtime, and other dependencies. In Docker, all dependencies that are needed to run the service, independent of the machine that is running it, are included. Containers are portable, lightweight, and can be deployed across various environments without worrying about compatibility issues. For readers who may be less familiar with containers, I should mention that containers are like virtual machines but are lightweight. Containerization has several benefits that we discuss here.

> **Portability**: Docker containers can run on various systems that support Docker. This makes it easy for developers to move the project from development to production.
>
> **Isolation**: Each container runs in isolation. This ensures that the dependencies of one service do not interfere with others. It is highly important when various codes are using different versions of some tools and libraries.
>
> **Scalability**: Containers can be scaled down or up easily, depending on the resource requirements of the microservice.

In addition, there are some other benefits, including easy and fast installation. From my experience, Docker has been one of the most valuable tools in building reliable ML systems. If some of these concepts feel new, don't worry—we will cover containerization in detail in Chapter 6. For now, the goal is simply to introduce the idea.

Virtual Environments

Virtual environments (e.g., **venv** or **Conda**) are commonly used in Python-based microservices for managing dependencies. A virtual environment makes an isolated Python environment with its own set of libraries. This ensures that various versions of dependencies of several code sets can coexist without conflicts. The following code shows how to create a virtual environment.

Listing 4-11. Creating a Virtual Environment

```
# Create a virtual environment
python -m venv myenv

# Activate the virtual environment
source myenv/bin/activate

# Install dependencies
pip install numpy pandas scikit-learn
```

Virtual environments are especially useful during the development phase. They allow developers to experiment with various versions of libraries without affecting other running services. If developers do not use a virtual environment, their other running codes may stop working without any clear reason.

Package Managers and Dependency Files

Package managers such as `pip` (for Python) or `npm` (for JavaScript) help in efficiently managing dependencies. In these tools, files like `requirements.txt` or `package.json` are used to specify the exact versions of libraries needed by a microservice. This ensures that all instances of the service use the same versions of dependencies. Package managers make the system reproducible and transferable. The example below shows a typical requirements.txt file.

Listing 4-12. The Contents of a Typical Requirement File

```
numpy==1.21.0
pandas==1.3.0
scikit-learn==0.24.2
```

Orchestration Tools

Kubernetes and other orchestration tools help manage and operate containerized microservices at scale. While Kubernetes does not manage dependencies directly, it ensures that the correct container images, each already packaged with their required dependencies, are deployed consistently. Also, Kubernetes helps with load balancing, managing resource allocation, and monitoring. These features make managing dependencies at scale much easier.

Helm Charts are often used in Kubernetes to package Kubernetes resources. These packages include container images, configuration files, and dependencies. To give you a better sense of orchestration. Assume you have several containers, you need someone to check them if they are running, and if they have crashed, they should restart them. Additionally, if the load on some containers increases, that person should allocate more resources to that container and should reduce the resources that are allocated to the containers with lower load. Do not worry, orchestration tools such as Kubernetes take care of all these tasks that I mentioned. Similar to containerization, we would have an extensive talk about orchestration in Chapter 6.

Challenges in Managing Dependencies

Here, we discuss the challenges in managing the dependencies, and in the next section, we provide some solutions to overcome these challenges.

> **Version Conflicts**: Various microservices may need different versions of the same library, leading to version conflicts. This is especially challenging in shared environments where multiple microservices are deployed on the same infrastructure or machine. Before explaining how we can overcome these problems, you can test yourself by answering how we can prevent version conflicts.
>
> **Dependency Drift**: Over time, the versions of dependencies or libraries that are used by various services may change. This leads to inconsistency or even crashes across the system. This is known as **dependency drift** and can cause issues such as incompatibility between microservices or unexpected behavior.
>
> **Complex Dependency Trees**: ML systems often rely on many libraries, each with its own dependencies. This creates deep, complex dependency trees. In these trees, multiple versions of the same library may be required, making dependency management difficult.

For example, imagine you need library A, which depends on version 1 of library B, but you also need library C, which depends on version 2 of library B. Managing such conflicts becomes challenging.

Note that here we are referring to dependencies within a single service or container.

Best Practices for Overcoming Dependency Challenges

In this section, we will review the best practices for overcoming dependency challenges. Also, you can find the answers to some of your questions or the questions that I have asked earlier.

> **Immutable Infrastructure**: Immutable infrastructure is a principle where you do not modify dependencies inside a running environment. Instead of updating libraries or configurations in place, you rebuild the container with updated dependencies and redeploy it. This approach minimizes dependency drift and ensures consistent environments.[24]
>
> **Dependency Scanning and Security**: Tools like `Snyk` and `Dependabot` can automatically scan for vulnerable or outdated dependencies and suggest updates. This ensures that all dependencies are secure and up to date.
>
> **Dependency Versioning and Lock Files**: Lock files, such as `package-lock.json` **or** `Pipfile.lock`, help in maintaining consistent versions of dependencies across different environments. These files lock the specific versions of libraries to be used, preventing inconsistencies during deployment.

Now you can see how the problem of complex dependency trees which was discussed earlier can be controlled using versioning, containerization, and immutability principles. Table 4-4 shows various tools and techniques that can be used for effective dependency management in ML microservices.

***Table 4-4.** Techniques for Managing Dependencies in Microservices-Based ML Systems*

Technique	Description	Tools and Technologies
Containerization	Use of containers to encapsulate dependencies	Podman, Docker
Virtual Environments	Isolated Python environments for dependency management	Conda, venv
Package Managers	Manage and install dependencies	pip, yarn, npm
Orchestration Tools	Manage and deploy containerized services	Helm Charts, Kubernetes
Dependency Scanning Tools	Automate the detection of vulnerable and outdated dependencies	Dependabot, Snyk

Before we finish the chapter with the project, I am going to tell you a point about containerization. Remember that the container is a different environment from your computer! In one project, a junior developer complained that a service worked correctly on his machine but failed when run inside Docker. After inspection, we discovered that the container did not have internet access, so it could not read data from the external API—whereas his local machine could. This is a common issue when developers assume containers inherit the same environment as their local system.

Now, as with previous chapters, we will finish with a project to help you apply what you have learned to a practical problem.

Project Title: Modular and Scalable ML Pipeline

Objective

The objective of this project is to build a scalable and modular machine learning (ML) pipeline based on what you have learned in this chapter. In this project, we consider the best practices such as high cohesion/low coupling, reusability, separation of concerns, and abstraction. By completing this project, you will learn how to design a real-world ML system using a microservices-inspired architecture. You will gain and test your practical experience in loading data, preprocessing, training a model, and performing inference—all within a scalable and maintainable structure.

Project Description

In this project, you will build a simple ML pipeline using Python. The pipeline is divided into several modules, each encapsulated within its own class. These modules are

- **Data Loading**: Responsible for obtaining or reading the dataset
- **Preprocessing**: Handles operations such as scaling and preparing the data
- **Model Training**: Trains a model using the processed data
- **Model Prediction**: Performs inference using the trained model

Pipeline Orchestration: Integrates all components and creates the complete end-to-end workflow. This design helps you to test your understanding of the content of this chapter in the modular design of an ML context. In addition, this project provides a clear, step-by-step approach for building scalable ML systems. This project focuses on modularity. It is recommended to learn modularity well and consider that in all your projects.

Step 1: Import Required Libraries

The code below imports the libraries that we need for this project. Before each project, ensure that you have installed all required libraries.

In this step, we import the required libraries for the project. These libraries include

- `NumPy` **and** `Pandas`**:** Essential for data processing and numerical computations
- `scikit-learn` Modules: These modules contain several functions and classes that have different functionalities
    - `load_iris` to load the Iris dataset (the dataset that is used in this project)
    - `train_test_split` to divide our dataset into training and testing subsets
    - `StandardScaler` for scaling features
    - `LogisticRegression` for training a logistic regression model
    - `accuracy_score` to evaluate model performance

These libraries form the main parts of our ML pipeline for correct processing and effective building of the model.

Listing 4-13. Step 1: Import Required Libraries

```
import numpy as np
import pandas as pd
from sklearn.datasets import load_iris
from sklearn.model_selection import train_test_split
from sklearn.preprocessing import StandardScaler
from sklearn.linear_model import LogisticRegression
from sklearn.metrics import accuracy_score
```

Expected Results

As you run the code of this step, no output will be generated. So, no output is good news that everything is fine and no error has happened.

Step 2: Data Loader Module

This step defines the `DataLoader` class, which encapsulates the process of fetching our dataset. In some of the previous projects, we were loading data directly; however, in this project, we want to practice modular design. When an instance of this class is initiated, it prints an initialization message. The `load_data` method loads the Iris dataset using scikit-learn's `load_iris()` function (we explained this function in the previous step). Then, it prints the number of samples and features in the dataset and returns both the feature matrix `X` and the label vector `y`.

Listing 4-14. Step 2: Data Loader Module

```
from sklearn.datasets import load_iris
import numpy as np

class DataLoader:

  def __init__(self):
    print("DataLoader initialized.")

  def load_data(self):
    """
```

```
    Load and return the Iris dataset.

    Returns:
      X (np.ndarray): Features.
      Y (np.ndarray): Labels.
    """
    print("Loading Iris dataset.")
    data = load_iris()
    X, y = data.data, data.target
    print(f"Dataset loaded with {X.shape[0]} samples and {X.shape[1]}
    features.")
    return X, y

# ---- Call the function ----
loader = DataLoader()
X, y = loader.load_data()
```

Expected Results

This step has some outputs, so no output is not good anymore. When you create a `DataLoader` instance and call `load_data()`, you should see the following outputs:

```
"DataLoader initialized."
"Loading Iris dataset."
"Dataset loaded with 150 samples and 4 features."
```

Step 3: Preprocessor Module

In this step, you write the code for the Preprocessor class. This class is responsible for scaling the dataset. The class uses StandardScaler to standardize the features by removing the mean and scaling to unit variance. Standardizing is a common statistical feature that helps to bring various data to one comparable scale. The fit_transform method is used on the training data to both fit the scaler and transform the data. For the test data, only the transform method is used. It is because the scaler has already been fitted on the training set.

In this step, we ensure that our preprocessing is applied consistently across training and testing sets.

Listing 4-15. Step 3: Preprocessor Module

```
class Preprocessor:
    def __init__(self):
        # Initialize the scaler.
        Self.scaler = StandardScaler()
        print("Preprocessor initialized with StandardScaler.")

    def fit_transform(self, X_train):
        """
        Fit the scaler on the training data and transform it.
        Args:
            X_train (np.ndarray): Training features.
        Returns:
            X_scaled (np.ndarray): Scaled training features.
        """
        print("Fitting and transforming training data.")
        X_scaled = self.scaler.fit_transform(X_train)
        return X_scaled

    def transform(self, X_test):
        """
        Transform the test data using the already fitted scaler.
        Args:
            X_test (np.ndarray): Test features.
        Returns:
            X_scaled (np.ndarray): Scaled test features.
        """
        print("Transforming test data.")
        return self.scaler.transform(X_test)

import numpy as np

# Example dummy data
X_train = np.array([[1.0, 2.0],
          [2.0, 3.0],
          [3.0, 4.0]])
```

```
X_test = np.array([[1.5, 2.5],
          [2.5, 3.5]])

# Call the Preprocessor class
pre = Preprocessor()

# Fit on training data and transform
X_train_scaled = pre.fit_transform(X_train)
print("Scaled X_train:\n", X_train_scaled)

# Transform test data
X_test_scaled = pre.transform(X_test)
print("Scaled X_test:\n", X_test_scaled)
```

Expected Results

Upon executing the preprocessing steps:

```
"Preprocessor initialized with StandardScaler."
"Fitting and transforming training data." (for training data)
"Transforming test data." (for test data)
```

Step 4: Model Trainer Module

This class represents the ML model training component. We name the class of this step `ModelTrainer`. The ModelTrainer class handles the training of our machine learning model. Here, we use logistic regression for classification tasks. The maximum number of iterations is set to 200 to ensure convergence; however, you can set that to various numbers for practicing. When the train method is called, it fits the logistic regression model to the scaled training data and prints status messages before and after training. So, this step has output too. The trained model can be retrieved using the get_model method.

Listing 4-16. Step 4: Model Trainer Module

```
class ModelTrainer:
    def __init__(self):
        # Using a simple logistic regression for demonstration.
        Self.model = LogisticRegression(max_iter=200)
        print("ModelTrainer initialized with LogisticRegression.")
```

```
    def train(self, X_train, y_train):
        """
        Train the model on the training data.
        Args:
            X_train (np.ndarray): Processed training features.
            Y_train (np.ndarray): Training labels.
        """
        print("Training model.")
        self.model.fit(X_train, y_train)
        print("Model training complete.")

    def get_model(self):
        """
        Return the trained model.
        """
        return self.model

# Dummy example data
X_train = np.array([[0.1, 0.5],
          [0.2, 0.4],
          [0.3, 0.3]])
y_train = np.array([0, 1, 0])

trainer = ModelTrainer()
trainer.train(X_train, y_train)
model = trainer.get_model()

print("Trained Model:", model)
```

Expected Results

After invoking the training process:

```
"ModelTrainer initialized with LogisticRegression."
"Training model." Followed by "Model training complete."
```

Step 5: Model Predictor Module

This module handles inference using the trained model. Now the model is trained, and we need a code for inference. We made a class and named it `ModelPredictor`. The `ModelPredictor` class is designed for the inference stage of our pipeline. It takes the trained model as input. The `predict` method applies the model to the test data and prints a message to indicate that predictions are being made. This modular approach allows the inference component to be maintained or scaled independently of the training process.

Listing 4-17. Step 5: Model Predictor Module

```
class ModelPredictor:
    def __init__(self, model):
        """
        Initialize with a trained model.
        Args:
            model: A trained machine learning model.
        """
        self.model = model
        print("ModelPredictor initialized.")

    def predict(self, X_test):
        """
        Generate predictions for the test data.
        Args:
            X_test (np.ndarray): Processed test features.
        Returns:
            predictions (np.ndarray): Predicted labels.
        """
        print("Making predictions.")
        predictions = self.model.predict(X_test)
        return predictions

predictor = ModelPredictor(model)
preds = predictor.predict(X_test)
print("Predictions:", preds)
```

Expected Results

When predictions are made:

```
"ModelPredictor initialized."
"Making predictions."
```

Step 6: Main Pipeline Orchestration

The code below orchestrates the pipelines.

Listing 4-18. Step 6: Main Pipeline Orchestration

```
def main():
    print("Starting ML pipeline orchestration.")

    # Step 6.1: Load Data
    data_loader = DataLoader()
    X, y = data_loader.load_data()

    # Step 6.2: Split Data into Training and Testing Sets
    print("Splitting data into train and test sets.")
    X_train, X_test, y_train, y_test = train_test_split(X, y, test_
    size=0.2, random_state=42)

    # Step 6.3: Preprocess Data
    preprocessor = Preprocessor()
    X_train_scaled = preprocessor.fit_transform(X_train)
    X_test_scaled = preprocessor.transform(X_test)

    # Step 6.4: Train the Model
    model_trainer = ModelTrainer()
    model_trainer.train(X_train_scaled, y_train)

    # Step 6.5: Inference - Make Predictions
    predictor = ModelPredictor(model_trainer.get_model())
    predictions = predictor.predict(X_test_scaled)

    # Step 6.6: Evaluate the Model
    accuracy = accuracy_score(y_test, predictions)
    print(f"Model Accuracy: {accuracy * 100:.2f}%")
```

This is the main step where we combine all the classes that we have made in `main`.

Step 7: Running the Main Pipeline

This ensures the main pipeline runs if the script is executed. The code below shows the script. The **Main Pipeline Orchestration** function, `main()`, brings together all the modules defined earlier. Here's what happens in each sub-step:

`Load Data`: An instance of `DataLoader` is created and used to load the Iris dataset.

`Split Data`: The dataset is split into training and testing sets using `train_test_split`.

`Preprocess Data`: The `Preprocessor` class is used to scale the training and test data.

`Train the Model`: The `ModelTrainer` class trains a logistic regression model on the scaled training data.

`Inference`: The `ModelPredictor` class generates predictions on the test data.

`Evaluation`: The model's performance is evaluated using accuracy, and the result is printed.

Each of these steps produces console output (using `print`) so that you can follow the progress of the pipeline. The modular design makes debugging and tracking easy.

Listing 4-19. Step 7: Running the Main Pipeline

```
if __name__ == "__main__":
    main()
```

Expected Results

When you run the `main` function, you will see the output of each stage of the pipeline:

- Initialization messages for each module
- Confirmation messages for each step of data loading, preprocessing, model training, and prediction
- A final printed statement illustrating the model accuracy, similar to above

```
Starting ML pipeline orchestration.
DataLoader initialized.
Loading Iris dataset.
Dataset loaded with 150 samples and 4 features.
Splitting data into train and test sets.
Preprocessor initialized with StandardScaler.
Fitting and transforming training data.
Transforming test data.
ModelTrainer initialized with LogisticRegression.
Training model.
Model training complete.
ModelPredictor initialized.
Making predictions.
Model Accuracy: 100.00%
```

Conclusion

This project demonstrates how to build a scalable, modular ML system by separating the workflow into distinct modules such as data loading, preprocessing, training, and prediction. Each module is encapsulated in its own class to ensure high cohesion, low coupling, and clear separation of concerns.

These principles make the code easier to maintain, reuse, and scale, especially when moving toward microservices. The chapter provides clear snippets, explanations, and expected outputs to guide you.

You can extend the project by creating API endpoints for each class, containerizing them with Docker, and structuring them as microservices.

Summary

Building modular and scalable ML systems requires careful planning around design patterns, dependency management, and cloud integration. A microservices architecture helps break ML workflows into smaller, independently deployable components, improving flexibility, scalability, and maintainability. However, it also introduces challenges such as dependency drift, version conflicts, inter-service communication overhead, and increased network load from API calls.

Managing dependencies is crucial for reproducibility and stable deployment. Tools like virtual environments, Docker, package managers, and orchestration frameworks help control versions and reduce conflicts. By following best practices and automation, teams can keep ML systems maintainable and scalable over time.

CHAPTER 5

Infrastructure for Machine Learning Workloads

As machine learning across industries grows, the choice of infrastructure becomes increasingly important. Infrastructure determines model scalability, performance, and cost-effectiveness. Machine learning workloads need significant storage, compute power, and data management capabilities.[25] Thus, selecting suitable infrastructure is complex and very impactful. Organizations must decide to deploy ML models in the cloud, on-premise, or in a hybrid approach. By on-premise, we refer to infrastructure hosted and managed within the organization's own environment (e.g., in-house servers or local data centers). For this purpose, they should consider factors such as security, cost, and performance.

In this chapter, we discuss the key infrastructure parameters for machine learning workloads. We start with the basic selection between on-premise and cloud-based ML infrastructure. Then, we will discuss the role of specialized hardware accelerators, including GPUs and TPUs, in enhancing ML efficiency. Finally, we provide cost optimization strategies to help organizations balance performance and expenses when deploying machine learning solutions at scale. I should mention that cost optimization is very critical in infrastructure selection. Of course, if the cost is not limited, selecting infrastructure will not be the matter; anyone could build a data center with a billion dollars' worth of hardware to run models at maximum speed. Realistically, however, engineering decisions must balance capability with financial constraints.

At the end of this chapter, you should have a solid understanding of ML infrastructure selection, hardware accelerators, and best practices for optimizing and reducing costs. This chapter enables you to handle scaling and deploy ML workloads effectively. We start by comparing cloud vs. on-premise ML infrastructure. If you have not read Chapter 4, I recommend reading it before reading the content of this chapter.

M. R. Mahdiani, *Mastering Machine Learning Architecture and Solutions*,
https://doi.org/10.1007/979-8-8688-2527-9_5

Cloud vs. On-Premise ML Infrastructure

Companies must decide between on-premise and cloud-based solutions for their ML system. This decision affects compute efficiency, data governance, business agility, and goals.

Cloud-based infrastructure offers scalability, flexibility, and access to cutting-edge hardware without requiring investment in physical infrastructure such as servers and specialized accelerators. On the other hand, on-premise infrastructure provides full control over computation and data, supporting strong privacy, security, and compliance requirements supporting strong privacy, security, and compliance requirements particularly important in regulated industries. For many organizations, a hybrid approach combining the benefits of each approach is the preferred solution.

In this section, we discuss the main parameters affecting infrastructure selection, including cost considerations, scalability, compliance, latency, and operational complexity. By understanding the advantages and disadvantages of each approach, you can make thoughtful decisions that align with your business goals and requirements. Before comparing these approaches, let's talk about each in more detail. We start with cloud-based, then on-premise, and finally, we talk about the hybrid approach.

Cloud-Based ML Infrastructure

In cloud-based infrastructure, we use cloud service providers (CSPs) such as Microsoft Azure, Google Cloud, AWS, and IBM Cloud to manage ML workloads. These cloud providers, especially the first three, are considered the leading cloud platforms in the industry. They offer storage, compute, and AI services with minimal upfront investment. These features make CSPs an attractive choice for businesses that are going to scale quickly.[26]

One of the primary benefits of cloud-based ML infrastructure is scalability. Cloud environments provide on-demand scaling of compute resources, which allows ML workloads to adjust dynamically based on demand. This flexibility is especially useful when training large-scale models or running inference on unpredictable workloads. Most companies, especially small startups that I have visited, prefer cloud services because they can get started quickly without managing hardware. However, I also encountered a finance-focused company that chose to operate its own servers due to strict data governance requirements. Cost efficiency is another key benefit. Instead of buying expensive hardware, cloud computing follows a pay-as-you-go model, where organizations only pay for the resources they use. In other words, instead of buying computational power, they just rent it. This shifts capital expenditure (CapEx) costs to operational expenditures (OpEx), improving financial flexibility. A simple analogy is comparing the cost of owning a luxury home to staying in a hotel—paying only when you need the service. Another benefit of using cloud provider services is that they offer fully managed ML services. These services help organizations avoid the overhead of infrastructure setup, configuration, and maintenance. Services such as Google Cloud Vertex AI, AWS SageMaker, and Azure ML Studio provide a smooth, streamlined workflow for data scientists and engineers. This approach helps users to focus on model development rather than hardware provisioning and maintenance.

Another main advantage of cloud-based ML infrastructure that is advertised in different texts is global accessibility. Cloud services allow distributed teams to access data, run tests, and deploy models from anywhere in the world, fostering collaboration across geographies. Remote work became very common after the pandemic, and cloud providers make it easy for remote-first teams. However, accessibility can vary across regions depending on internet availability or local restrictions. On the other hand, on-premise systems can also support remote teams if appropriate networking solutions are implemented—I have visited several companies that use on-premise infrastructure with fully remote ML teams.

Security and compliance are also crucial. Leading CSPs comply with industry standards such as GDPR, ISO 27001, HIPAA, and SOC 2. CSPs provide built-in security measures such as identity and access management (IAM) policies, encryption, and secure network configurations. Of course, these protections are only effective if software teams follow best practices and avoid introducing vulnerabilities.

Despite these benefits, cloud-based ML infrastructure presents some challenges. Recurring costs may increase over time, particularly for compute-heavy workloads such as deep learning training. We may have latency issues due to remote data access. Additionally, vendor lock-in is another issue, as migrating ML workflows from one CSP to another CSP is complex and difficult (if not impossible!).

On-Premise ML Infrastructure

On-premise ML infrastructure includes managing and deploying physical servers, storage systems, and networking components within the organization's own data center or hardware environment. Companies selecting on-premise ML infrastructure usually require performance optimization, strict data control, and compliance adherence.

One of the first advantages of on-premise ML infrastructure is security and data privacy. Organizations handling sensitive data, such as those in finance, healthcare, and government sectors, prefer on-premise infrastructure. By this, they keep full control over their data access, storage, and security policies. On-premise architecture ensures that third parties do not have access to company data.

Performance optimization is also another important benefit. On-premise hardware can be adjusted for high-performance ML workloads. They guarantee to minimize latency and ensure high-throughput data processing. This is particularly important in real-time cases such as autonomous systems, fraud detection, and industrial automation.

Another consideration is cost predictability. Unlike cloud-based services that incur recurring fees, on-premise infrastructure requires a one-time hardware investment. This is more cost-effective for organizations running continuous, high-demand ML workloads. In such cases, the ongoing costs are mainly electricity, networking, cooling, and hardware maintenance. However, these operational expenses can sometimes exceed the cost of using cloud service providers. Again, think of it like renting vs. buying a house: even after buying, you still pay for utilities, maintenance, and taxes. Compliance and regulatory control are the main factors affecting the choice of on-premise ML infrastructure. Industries governed by strict regulations, such as HIPAA, GDPR, SOC 2, and PCI DSS, often prefer on-premise data storage and processing. For these organizations, on-premise is better aligned with compliance with legal and industry standards.

On the other hand, on-premise ML infrastructure has disadvantages too. The initial price of hardware selection, procurement, and setup is high. Companies need experts to set up their systems. Maintenance is expensive. Scalability is limited, and expanding compute resources requires additional hardware procurement. Maintaining one premise architecture is time-consuming and costly. Additionally, organizations have to manage software updates and maintenance, hardware failures, and security patches, which increase their operational complexity. Companies may have more downtime and waste the time of their team. Additionally, these downtimes can affect their branding. What we described here was the worst case; do not panic in selecting on-premises or cloud services.

Hybrid ML Infrastructure: The Best of Both Worlds?

To balance performance, flexibility, and cost, some organizations use a hybrid ML infrastructure. For that, they combine cloud services with on-premise resources. This approach enables businesses to maintain sensitive data on-premise while using the cloud for compute-intensive tasks such as deep learning training and large-scale inference. However, if the hybrid architecture is poorly designed, organizations may face the challenges of both environments rather than benefiting from their strengths. A common hybrid ML use case is storing sensitive data on-premise while using cloud-based TPUs or GPUs for training complex ML models. For less familiar readers, a TPU (Tensor Processing Unit) is a hardware accelerator similar to a GPU but optimized specifically for large-scale machine learning workloads. This allows organizations to have data security while using the cloud's computational power. Keep in mind that they cannot keep all their data on-premise, and anyway, organizations must share some part of their data in the cloud. Another hybrid pattern is training models on-premise and deploying them in the cloud for inference, which helps optimize cost and latency depending on the use case. Some companies also design for "burst workloads," where on-premise resources handle routine operations while cloud instances are activated only during peak demand. Keep in mind that hybrid infrastructure requires maintaining and managing both cloud and on-premise systems, so it should be chosen carefully. Figure 5-1 illustrates a hybrid ML setup where data is stored on-premise while training and inference occur in the cloud.

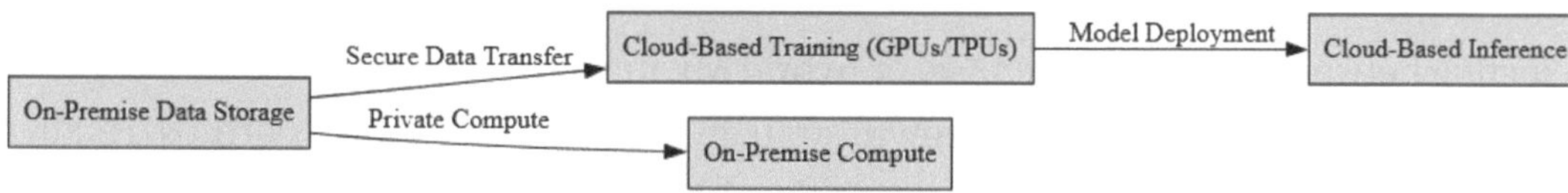

Figure 5-1. *Hybrid ML Infrastructure*

In this book, I try to maintain a fair perspective and present both the benefits and challenges of each approach. The final decision depends entirely on your organizational needs and constraints.

Cost Analysis: Cloud vs. On-Premise

Cost considerations play an important role in selecting ML infrastructure. Table 5-1 compares the cost of cloud and on-premise infrastructure.

Table 5-1. *Comparison of Cost Factors Between Cloud and On-Premise ML Infrastructure*

Cost Factor	Cloud ML Infrastructure	On-Premise ML Infrastructure
Initial Investment	Low (pay-per-use)	High (hardware and setup costs)
Scalability	High (elastic resources)	Limited (requires hardware procurement)
Operational Costs	Recurring cloud fees	Maintenance and personnel costs
Long-Term Viability	Cost-effective for variable workloads	Cost-effective for consistent workloads
Data Transfer Costs	Cloud egress fees apply	No additional costs

Cloud services offer low upfront costs but would be expensive over time, especially for high-volume compute workloads. In contrast, on-premise infrastructure needs a significant initial investment but offers more predictable long-term costs.

Making the Right Choice

Selecting between cloud, on-premise, or hybrid ML infrastructure depends on budget, workload, compliance needs, and growth strategy. Research teams and startups usually benefit from cloud-based ML solutions, as cloud-based solutions are more flexible with less upfront costs. On the other hand, enterprises with strict data security requirements may prefer on-premise or hybrid approaches. Finally, organizations with unpredictable workloads can optimize cost and performance by leveraging hybrid solutions.[27] In the next section, we discuss the hardware components in more detail.

Leveraging GPUs and TPUs

As machine learning models become bigger, more complex, and with more parameters, their computational demands increase. A decade ago, traditional central processing units (CPUs), were a good choice for many ML models. However, today, CPUs often struggle in efficiently handling large-scale machine learning (ML) workloads. This issue has led to the use of specialized hardware. These processors offer substantial speedups for deep learning and other parallelizable ML tasks.[28]

GPUs were originally designed for graphics rendering but are now widely used as parallel processors for ML due to their high throughput. TPUs, in contrast, are custom-built specifically for machine learning, particularly tensor operations used in deep learning. In this section, we examine the benefits, architectures, and real-world applications of GPUs and TPUs to show how they enable efficient ML workloads.

The Evolution of Machine Learning Hardware

Initial ML algorithms were predominantly trained on CPUs, which offer a fixed number of cores for sequential processing. As deep learning models got more sophisticated with millions or billions of parameters, the need for large-scale parallel computation increases significantly.

GPUs emerged as an effective alternative because they contain thousands of smaller cores capable of performing many operations in parallel—especially SIMD-style operations on large matrices. While CPUs excel at single-threaded, low-latency tasks, GPUs are better suited for high-throughput training and inference in deep learning.

To meet the demand for even greater ML performance, Google introduced TPUs—custom accelerators designed specifically for tensor operations and large matrix multiplications. TPUs provide optimizations such as systolic array architectures, enabling higher throughput and improved energy efficiency compared to general-purpose GPUs.

GPU Architecture and Its Role in ML

GPUs are highly effective for deep learning tasks that require processing large amounts of data in parallel. The main structure of a GPU is composed of

Streaming Multiprocessors (SMs): Every GPU is made up of many SMs that each can contain hundreds of cores that can work with many threads simultaneously.

High-Bandwidth Memory (HBM): GPUs have high-speed memory for efficiently handling large tensors and doing intermediate computations.

CUDA and Tensor Cores: NVIDIA GPUs support CUDA (Compute Unified Device Architecture) for general-purpose computing and Tensor Cores that accelerate mixed-precision matrix operations used in deep learning.

These architectural features make GPUs very efficient for operations such as convolutions, matrix multiplications, and activation computations—key components of deep learning workloads. Now, in the following Python code, we can see how GPUs accelerate ML workloads using TensorFlow.

Listing 5-1. Using TensorFlow and GPU for Handling Workload

```
import tensorflow as tf
import time

# Create a simple matrix multiplication workload
A = tf.random.normal([1000, 1000])
B = tf.random.normal([1000, 1000])

# Run on CPU
with tf.device('/CPU:0'):
    start_time = time.time()
    result_cpu = tf.matmul(A, B)
    print("CPU Execution Time:", time.time() - start_time)

# Run on GPU
if tf.config.list_physical_devices('GPU'):
```

```
    with tf.device('/GPU:0'):
        start_time = time.time()
        result_gpu = tf.matmul(A, B)
        print("GPU Execution Time:", time.time() - start_time)
else:
    print("No GPU detected.")
```

Expected Results

```
CPU Execution Time: 0.3240845203399658
GPU Execution Time: 0.18538212776184082
```

The above code is not for ML training or inferring; it just compares the speed of CPU and GPU for matrix multiplication. You can run that on your computer and see the difference (if your computer has a GPU and supports CUDA).

TPU Architecture and Deep Learning Optimization

TPUs are specialized accelerators designed specifically for tensor operations, not general-purpose computing. Because of this specialization, TPUs are highly effective for machine learning workloads.

The main components and characteristics of TPUs include

Matrix Multiplication Units (MXUs): TPUs include dedicated MXUs that accelerate large matrix multiplications—the core operation in deep learning—significantly reducing training and inference time.

High-Speed Interconnects: TPUs can be linked together into TPU Pods, allowing many TPUs to work collaboratively for large-scale distributed ML training.

Lower Power Consumption: TPUs are optimized for high performance per watt, often consuming less energy than GPUs for comparable deep learning workloads.

Using TPUs in TensorFlow is straightforward. The example below demonstrates how TensorFlow detects and assigns a TPU automatically.

Listing 5-2. Using TensorFlow and TPU for Handling Workload

```
import tensorflow as tf

# Detect and assign TPU if available
try:
    resolver = tf.distribute.cluster_resolver.TPUClusterResolver()
    tf.config.experimental_connect_to_cluster(resolver)
    tf.tpu.experimental.initialize_tpu_system(resolver)
    strategy = tf.distribute.TPUStrategy(resolver)
except ValueError:
    strategy = tf.distribute.MirroredStrategy()  # Fall back to GPU/CPU if
    TPU unavailable

print("Using strategy:", strategy)
```

Expected Results

In the above code, if a TPU is not available, TensorFlow gracefully falls back to a GPU or CPU.

```
INFO:tensorflow:Using MirroredStrategy with devices ('/job:localhost/
replica:0/task:0/device:CPU:0',)
Using strategy: <tensorflow.python.distribute.mirrored_strategy.
MirroredStrategy object at 0x000001EED9F90AD0>
```

Performance Comparison: GPU vs. TPU

Workload, type, pricing, and scalability, among other considerations, would determine a head-to-head comparison of GPUs and TPUs. Table 5-2 illustrates an overview of key differences between GPUs and TPUs.

Table 5-2. *Comparison Between GPUs and TPUs*

Feature	GPUs	TPUs
Optimization For	General-purpose computing	Deep learning tensor ops
Parallelism	High	Extremely high
Memory Bandwidth	High	Higher (optimized for tensors)
Power Consumption	Higher	Lower
Scalability	Scales with multiple GPUs	Scales with TPU Pods
Best Used For	Training and inference	Large-scale deep learning

This comparison emphasizes how GPUs are suitable for different workloads, while TPUs provide specialized advantages for deep learning. Figure 5-2 visually represents the architecture of both GPUs and TPUs.

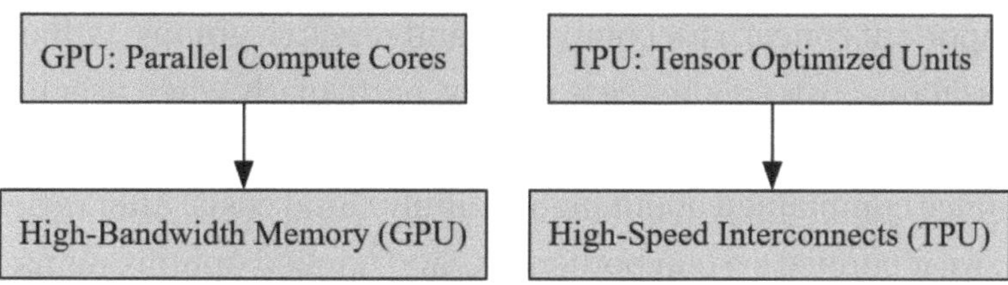

Figure 5-2. *Architecture of GPUs and TPUs*

When to Use GPUs vs. TPUs

The selection between GPUs and TPUs can be done based on specific ML workload requirements:

- **GPUs are good for**
    - General-purpose deep learning training and inference
    - Developing a custom ML model
    - When high memory bandwidth is needed

- **TPUs are suitable for**
  - Tasks including large-scale TensorFlow deep learning
  - Cloud-based ML systems
  - When energy consumption is important

For most projects, GPUs are the preferred option because they support multiple ML frameworks and are more flexible. However, TPUs provide significant advantages for organizations using Google Cloud or training very large-scale deep learning models.

Cost Optimization for ML Systems

Machine learning workloads are computationally expensive, which requires considerable resources for data processing, model training, and inference. As models get more complicated and datasets grow in volume, managing infrastructure expenses becomes an important concern for companies and research groups. Without thoughtful expense optimization, costs can increase rapidly, particularly when user utilizes cloud-based resources. For this reason, cost optimization is very important. For example, one company once complained about high monthly cloud costs. After reviewing their infrastructure, we identified system bottlenecks and applied optimizations, reducing their monthly expenses by more than half. This illustrates the critical importance of cost optimization. Without it, organizations risk becoming overly dependent on cloud providers, paying disproportionately for services.

Cost optimization in ML systems includes decisions about selecting infrastructure, allocating resources, scheduling workload, model efficiency, and best practices. In this section, we explore the main cost-saving techniques for ML workloads in storage, compute, data transfer, and model deployment. It should be considered that with cost optimization, we are not going to trade off the performance and reliability of the system. We just want to cut excessive costs.

Cost Components in ML Workloads

Before considering cost-saving measures, it is crucial to understand the main cost drivers in your machine learning infrastructure. Below, we outline each cost component and highlight key considerations for managing costs. The main cost components are

Compute Costs: TPUs and GPUs (or high-performance CPUs) are required for compute-intensive workloads such as model training. Cloud compute instances could be priced as reserved, on-demand, or spot instances, which has a great effect on cost.

Storage Costs: Huge datasets that are used for training and inference result in high storage costs, specifically in cloud environments. Different storage classes (e.g., cold storage vs. hot storage) determine the pricing.

Data Transfer Costs: Transferring big datasets between cloud services causes egress costs (it will be charged whenever data leaves the cloud provider's network). Hybrid cloud architectures and cross-region transfers can increase expenses.

Inference Costs and Model Serving: Deploying models for real-time inference needs steady computing resources, which can be costly if not optimized. Batch inference, which processes predictions at scheduled intervals, can help reduce costs.

Table 5-3 illustrates common ML infrastructure cost components and their associated pricing models.

Table 5-3. *Common ML Infrastructure Cost*

Cost Component	Description	Pricing Model
Compute	CPU, GPU, or TPU usage	On-demand, reserved, spot instances
Storage	Dataset storage for training and inference	Hot, cold, archival storage
Data Transfer	Moving data within or across cloud providers	Ingress (free), egress (charged)
Inference	Model serving infrastructure	Real-time vs. batch inference

Optimizing Compute Costs

Compute resources are one of the biggest expenses in ML workloads. Some strategies can help decrease the computational costs without decreasing the performance:

Selecting the Right Instance Type: Cloud providers offer different instance types optimized for various workloads. **GPUs/TPUs** are necessary for deep learning training; however, for smaller workloads, high-performance CPUs may be enough.

Using Spot and Preemptible Instances: Cloud providers provide **spot instances (AWS)** and **preemptible VMs (GCP)** at considerably reduced prices compared to on-demand instances. These instances are perfect for fault-tolerant workloads like model training.

Autoscaling and On-Demand Provisioning: By adjusting resources dynamically based on workload demand, you can prevent over-provisioning. **Kubernetes-based autoscaling** helps allocate resources efficiently in ML model deployment.

The following Python code demonstrates how to select the best GPU instance on AWS using Boto3, an AWS SDK for Python.

Listing 5-3. An Example of Using Boto3

```
import boto3

# Define the AWS region explicitly
# NOTE: Replace 'us-east-1' with the region you want to query.
REGION = 'us-east-1'

# Initialize EC2 client with the specified region
ec2 = boto3.client('ec2', region_name=REGION)

# Retrieve available GPU instance types
try:
    gpu_instances = ec2.describe_instance_types(
        Filters=[{'Name': 'processor-info.supported-gpus', 'Values':
        ['NVIDIA']}]
    )
```

```
    # Print instance details
    for instance in gpu_instances['InstanceTypes']:
        instance_type = instance.get('InstanceType', 'N/A')
        vcpus = instance.get('VCpuInfo', {}).get('DefaultVCpus', 'N/A')
        memory_mib = instance.get('MemoryInfo', {}).get('SizeInMiB', 'N/A')

        print(f"Instance Type: {instance_type}, vCPUs: {vcpus}, Memory:
        {memory_mib} MiB")

except Exception as e:
    # Handle potential exceptions like invalid region or
    authorization issues
    print(f"An error occurred: {e}")
```

Expected Results

You have to set AWS credentials before running your code; if not, you will see the following error.

```
An error occurred: Unable to locate credentials
```

This script gets a list of GPU instances on AWS and allows users to select the most cost-effective option for their ML workloads. In the next parts, we will talk about data costs. Based on your configuration, you have a different output.

Optimizing Storage Costs

Storage costs can increase quickly due to storage accumulation, specifically while working with large datasets for ML inference and training. Main storage optimization methods include

Choosing a Suitable Storage Tier: Hot storage is perfect for frequently accessed data; however, **cold storage** (e.g., AWS S3 Glacier, Google Coldline) is cheaper for archiving.

Data Compression and Format Selection: Using compressed data formats like **Parquet or Avro** decreases storage costs and improves read performance.

Object Lifecycle Policies: By automating data retention policies, you can ensure that non-required data is deleted or it has been moved to a lower-cost storage tier.

Sharding and Partitioning Large Datasets: Partitioning data by time (mostly) or category minimizes the costs of queries in data-intensive ML workloads.

Managing Data Transfer Costs

Cloud providers usually charge for data egress when data is transferred outside the cloud provider's network. Here, we offer some strategies to reduce these costs:

Keeping Compute and Data in the Same Region: By running ML workloads in the same region where the data is stored, you can prevent the costs related to cross-region transfer.

Using Content Delivery Networks (CDNs): Distributing ML models in such a way that they would be closer to end users reduces both egress costs and inference latency.

Batching API Requests: Instead of separate processing of each inference request, batching several requests reduces API call overhead.

Optimizing Model Training and Inference

Deployment strategies and efficient model training can considerably decrease the costs. Do not underestimate optimizations. If you can optimize your model to use fewer resources while not sacrificing accuracy and performance, you are the winner. In this section, we offer some best practices for optimization. In addition to these techniques, be innovative and invent your own optimization techniques.

Model Pruning and Quantization: Decreasing model size reduces compute and memory requirements, leading to cheaper inference costs. One technique to reduce model size is quantization. Quantization converts floating-point models to a lower precision format (e.g., INT8 instead of FP32), improving efficiency.

Using Distillation for Faster and Smaller Models: Knowledge distillation is a method that moves knowledge from large models to smaller ones. Distillation reduces computational costs while keeping accuracy.

Batch vs. Real-Time Inference: Batch inference handles and analyzes several inputs at once, decreasing the need for expensive real-time compute resources.

The code below shows how model quantization can decrease the inference costs using TensorFlow Lite.

Listing 5-4. Model Quantization Sample Code

```
import tensorflow as tf

# Load a pre-trained model
model = tf.keras.applications.MobileNetV2(weights="imagenet")

# Convert the model to TensorFlow Lite (TFLite) format
converter = tf.lite.TFLiteConverter.from_keras_model(model)
converter.optimizations = [tf.lite.Optimize.DEFAULT]
quantized_model = converter.convert()

# Save the quantized model
with open("model_quantized.tflite", "wb") as f:
    f.write(quantized_model)
```

Expected Results

```
Downloading data from https://storage.googleapis.com/tensorflow/keras-
applications/mobilenet_v2/mobilenet_v2_weights_tf_dim_ordering_tf_
kernels_1.0_224.h5
14536120/14536120 ———————————————————————————————— 2s 0us/step
Saved artifact at '/tmp/tmp86cpojwm'. The following endpoints are
available:

* Endpoint 'serve'
  args_0 (POSITIONAL_ONLY): TensorSpec(shape=(None, 224, 224, 3), dtype=tf.
  float32, name='keras_tensor')
```

```
Output Type:
  TensorSpec(shape=(None, 1000), dtype=tf.float32, name=None)
Captures:
  137802845167568: TensorSpec(shape=(), dtype=tf.resource, name=None)
  137802845168912: TensorSpec(shape=(), dtype=tf.resource, name=None)
  137802845171600: TensorSpec(shape=(), dtype=tf.resource, name=None)
```

The above code converts a deep learning model to a quantized format. This conversion decreases the compute requirements and memory usage for inference. In this chapter, we discussed various types of infrastructure, hardware, and techniques for cost optimization. Here, we finish the discussion of this chapter with a project.

Project Title: ML Infrastructure Cost and Performance Simulator

Objective

In the previous chapter's project, we were working on making an ML model. In this chapter, we talk about infrastructure.

This project helps you understand and compare various machine learning (ML) infrastructures, including on-premises, cloud, and hybrid. Additionally, here, we discuss the performance specification of different ML hardware (GPUs and TPUs). By simulating hardware performance and cost in the cloud, you will learn how to analyze cost optimization and make informed selections about your ML infrastructure. This project examines your learning of the concepts of this chapter on ML infrastructure, such as hardware utilization and cost analysis. Also, in this project, you can test your skills in designing scalable, modular simulations.

Project Description

In this project, you make a Python simulation that

- Models different ML infrastructure types (cloud, on-premise, hybrid) and calculates estimated costs based on storage, compute, and data transfer components

- Simulates the differences in performance between GPUs and TPUs for ML workloads
- Considers both the cost and performance simulations and tries to help you decide on the optimized cost solution for given workload parameters while keeping the performance

This project is designed using modular classes that separate concerns. In a way that one set of classes handles cost estimation (for different infrastructures) and the other set simulates the performance of hardware. We had a talk about this practice in Chapter 4. In each part, detailed explanations and output messages are provided.

Step 1: Import Required Libraries

In the first step, we import the necessary libraries for numerical computations and plotting. The below code imports the libraries. If you have not installed libraries in your environment, you have to install them first using `pip install numpy matplotlib`. In this step, we import `NumPy` for numerical analysis and `matplotlib` for plotting figures. These libraries are required to help us perform calculations and visualize simulation results.

Listing 5-5. Step 1: Import Required Libraries

```
import numpy as np
import matplotlib.pyplot as plt
```

Expected Results

This step should run without any output. If there is no output, it means there is no error, and the required libraries are loaded successfully.

Step 2: Define ML Infrastructure Cost Models

The code below simulates infrastructure with compute, storage, and data transfer cost components. It also implements cloud, on-premises, and hybrid versions by subclassing. In this code, after loading the required classes, we define classes for our cost model including:

MLInfrastructure: Is a base class for the common components of the ML infrastructure, including compute, storage, and data transfer. This class contains methods for computing the individual and total costs of the mentioned infrastructures.

CloudInfrastructure: Inherits from MLInfrastructure and uses some typical rates for cloud services. It adds a slight overhead for simulating additional cloud costs.

OnPremiseInfrastructure: This class uses lower compute rates, but it has higher storage and data transfer costs. Also, for simulating the cost of buying and maintaining hardware, it adds a fixed overhead.

HybridInfrastructure: Uses average rates from on-premise and cloud solutions and considers a small integration overhead.

Each class's total_cost() method computes the total estimated cost based on the input workload parameters.

Listing 5-6. Step 2: Define ML Infrastructure Cost Models

```
class MLInfrastructure:
    def __init__(self, compute_hours, compute_rate, storage_gb, storage_
    rate, data_transfer_gb, transfer_rate):
        """
        Initialize the MLInfrastructure with cost components.

        Args:
            compute_hours (float): Number of compute hours required.
            compute_rate (float): Cost per compute hour.
            storage_gb (float): Amount of storage required in GB.
            storage_rate (float): Cost per GB storage.
            data_transfer_gb (float): Data transfer volume in GB.
            transfer_rate (float): Cost per GB data transfer.
        """
        self.compute_hours = compute_hours
        self.compute_rate = compute_rate
        self.storage_gb = storage_gb
```

```
        self.storage_rate = storage_rate
        self.data_transfer_gb = data_transfer_gb
        self.transfer_rate = transfer_rate

    def compute_cost(self):
        # Cost for compute resources
        return self.compute_hours * self.compute_rate

    def storage_cost(self):
        # Cost for storage resources
        return self.storage_gb * self.storage_rate

    def data_transfer_cost(self):
        # Cost for data transfer resources
        return self.data_transfer_gb * self.transfer_rate

    def total_cost(self):
        # Total cost is the sum of compute, storage, and data
        transfer costs.
        return self.compute_cost() + self.storage_cost() + self.data_
        transfer_cost()

class CloudInfrastructure(MLInfrastructure):
    def __init__(self, compute_hours, storage_gb, data_transfer_gb):
        # Cloud cost rates can be assumed based on current market prices.
        # These rates are in dollars.
        compute_rate = 0.5      # $0.5 per compute hour
        storage_rate = 0.1      # $0.1 per GB storage
        transfer_rate = 0.12    # $0.12 per GB data transfer
        super().__init__(compute_hours, compute_rate, storage_gb, storage_
        rate, data_transfer_gb, transfer_rate)

    def total_cost(self):
        # Additional cloud overheads can be factored in if needed.
        overhead = 0.05 * super().total_cost()  # 5% overhead
        return super().total_cost() + overhead
```

```
class OnPremiseInfrastructure(MLInfrastructure):
    def __init__(self, compute_hours, storage_gb, data_transfer_gb):
        # On-Premise rates might be lower for compute but higher for
        storage and data management.
        compute_rate = 0.3      # $0.3 per compute hour (lower cost due to
        capital investment)
        storage_rate = 0.15     # $0.15 per GB storage
        transfer_rate = 0.08    # $0.08 per GB data transfer
        super().__init__(compute_hours, compute_rate, storage_gb, storage_
        rate, data_transfer_gb, transfer_rate)

    def total_cost(self):
        # For on-premise, consider amortized capital costs as a fixed
        overhead.
        fixed_overhead = 1000  # Fixed overhead cost for hardware
        investment
        return super().total_cost() + fixed_overhead

class HybridInfrastructure(MLInfrastructure):
    def __init__(self, compute_hours, storage_gb, data_transfer_gb):
        # Hybrid may combine cloud and on-premise; we use averaged rates.
        compute_rate = (0.5 + 0.3) / 2      # Average compute rate
        storage_rate = (0.1 + 0.15) / 2       # Average storage rate
        transfer_rate = (0.12 + 0.08) / 2     # Average transfer rate
        super().__init__(compute_hours, compute_rate, storage_gb, storage_
        rate, data_transfer_gb, transfer_rate)

    def total_cost(self):
        # Hybrid may incur integration overheads.
        integration_overhead = 0.03 * super().total_cost()  # 3% overhead
        return super().total_cost() + integration_overhead

#Above code does not have any output. If you want to test that, you can use
the following code:
compute_hours = 200         # hours of compute
storage_gb = 500            # storage in GB
data_transfer_gb = 300      # data transfer in GB
```

```
# Create infrastructure objects
cloud = CloudInfrastructure(compute_hours, storage_gb, data_transfer_gb)
on_prem = OnPremiseInfrastructure(compute_hours, storage_gb, data_
transfer_gb)
hybrid = HybridInfrastructure(compute_hours, storage_gb, data_transfer_gb)

# Print total cost for each
print("Cloud Infrastructure Cost:", cloud.total_cost())
print("On-Premise Infrastructure Cost:", on_prem.total_cost())
print("Hybrid Infrastructure Cost:", hybrid.total_cost())
```

Expected Results

This simulation, part of the above code, will produce the following results:

```
Cloud Infrastructure Cost: 195.3
On-Premise Infrastructure Cost: 1159.0
Hybrid Infrastructure Cost: 177.675
```

Step 3: Define Hardware Performance Simulator

The code below simulates the relative performance differences between GPU and TPU. In the below code, the `HardwareSimulator` class simulates the performance of training on different ML hardware. The `simulate_performance` method gets a hardware type (GPU or TPU) and the workload size; then it calculates the estimated training time by considering a performance factor. We have assigned a factor to GPUs and TPUs to simulate their performance. GPUs have a factor of 1.2, meaning they are a bit slower. TPUs have a factor of 0.8, meaning they are faster.

This simulation helps in getting the effect of hardware choice on ML workload performance.

Listing 5-7. Step 3: Define Hardware Performance Simulator

```
class HardwareSimulator:
    def __init__(self):
        print("Hardware Simulator initialized.")

    def simulate_performance(self, hardware_type, workload_size):
        """
```

```
        Simulate the performance (e.g., training time) of a given
        hardware type.

        Args:
            hardware_type (str): 'GPU' or 'TPU'.
            workload_size (float): A factor representing the workload size
            (e.g., data size or model complexity).

        Returns:
            float: Simulated training time (in hours).
        """
        base_time = workload_size  # Base time proportional to
        workload size
        if hardware_type.upper() == 'GPU':
            # Assume GPU takes base_time multiplied by a factor.
            performance_factor = 1.2  # GPU is slightly slower than TPU for
            this simulation
        elif hardware_type.upper() == 'TPU':
            performance_factor = 0.8  # TPU is faster
        else:
            raise ValueError("Invalid hardware type. Choose 'GPU' or
            'TPU'.")

        simulated_time = base_time * performance_factor
        print(f"Simulated training time using {hardware_type.upper()}:
        {simulated_time:.2f} hours.")
        return simulated_time

# To generate some output we can add the following code:
sim = HardwareSimulator()
sim.simulate_performance("GPU", 10)
sim.simulate_performance("TPU", 10)
```

Expected Results

By running the simulation for different hardware types, you will print the simulated training time. For instance:

```
Hardware Simulator initialized.
Simulated training time using GPU: 12.00 hours.
Simulated training time using TPU: 8.00 hours.
```

Step 4: Main Simulation Orchestration

The code demonstrates how to model cloud resources using a class and calculate their costs using an encapsulated method (total_cost()). In this code: The `main_simulation()` function runs the simulation by

- Defining workload parameters (storage, compute hours, data transfer, and workload size)
- Creating instances for all infrastructure model (on-premise, cloud, and hybrid) and printing their estimated costs
- Using the HardwareSimulator for calculating and printing simulated training times for GPUs and TPUs
- In addition, this step plots two charts:
    - A bar chart that compares the estimated costs of all three infrastructure types
    - A bar chart that compares the simulated training times for different hardware (GPU and TPU)

These outputs provide a complete view of both performance and cost, helping in decision-making for ML workloads.

Listing 5-8. Step 4: Main Simulation Orchestration

```
def main_simulation():
    print("Starting ML Infrastructure and Hardware Performance
    Simulation.\n")

    # Define workload parameters
    compute_hours = 500
    storage_gb = 200
    data_transfer_gb = 150
    workload_size = 10
```

```
print("Workload Parameters:")
print(f"Compute Hours: {compute_hours}")
print(f"Storage (GB): {storage_gb}")
print(f"Data Transfer (GB): {data_transfer_gb}")
print(f"Workload Size Factor: {workload_size}\n")

# Instantiate and compute cost for Cloud Infrastructure
cloud = CloudInfrastructure(compute_hours, storage_gb, data_
transfer_gb)
cloud_cost = cloud.total_cost()
print(f"Estimated Cloud Infrastructure Cost: ${cloud_cost:.2f}")

on_prem = OnPremiseInfrastructure(compute_hours, storage_gb, data_
transfer_gb)
on_prem_cost = on_prem.total_cost()
print(f"Estimated On-Premise Infrastructure Cost: ${on_prem_cost:.2f}")

hybrid = HybridInfrastructure(compute_hours, storage_gb, data_
transfer_gb)
hybrid_cost = hybrid.total_cost()
print(f"Estimated Hybrid Infrastructure Cost: ${hybrid_cost:.2f}\n")

hardware_sim = HardwareSimulator()
gpu_time = hardware_sim.simulate_performance("GPU", workload_size)
tpu_time = hardware_sim.simulate_performance("TPU", workload_size)

infrastructures = ['Cloud', 'On-Premise', 'Hybrid']
costs = [cloud_cost, on_prem_cost, hybrid_cost]

plt.figure(figsize=(8, 5))
plt.bar(infrastructures, costs, color='skyblue')
plt.xlabel('Infrastructure Type')
plt.ylabel('Estimated Cost ($)')
plt.title('Cost Comparison of ML Infrastructures')
plt.tight_layout()
plt.show()  # Figure 3: Cost Comparison Bar Chart

hardware = ['GPU', 'TPU']
times = [gpu_time, tpu_time]
```

```
    plt.figure(figsize=(8, 5))
    plt.bar(hardware, times, color='lightgreen')
    plt.xlabel('Hardware Type')
    plt.ylabel('Simulated Training Time (hours)')
    plt.title('Hardware Performance Comparison')
    plt.tight_layout()
    plt.show()  # Figure 4: Hardware Performance Comparison Bar Chart

if __name__ == "__main__":
    main_simulation()
```

Expected Results

```
Starting ML Infrastructure and Hardware Performance Simulation.

Workload Parameters:
Compute Hours: 500
Storage (GB): 200
Data Transfer (GB): 150
Workload Size Factor: 10

Estimated Cloud Infrastructure Cost: $302.40
Estimated On-Premise Infrastructure Cost: $1192.00
Estimated Hybrid Infrastructure Cost: $247.20

Hardware Simulator initialized.
Simulated training time using GPU: 12.00 hours.
Simulated training time using TPU: 8.00 hours.
```

After running main_simulation(), you should see the printed outputs of the workload parameters and the estimated costs for each infrastructure type. Additionally, you should see simulated training times for GPU and TPU (Figure 5-3 and Figure 5-4).

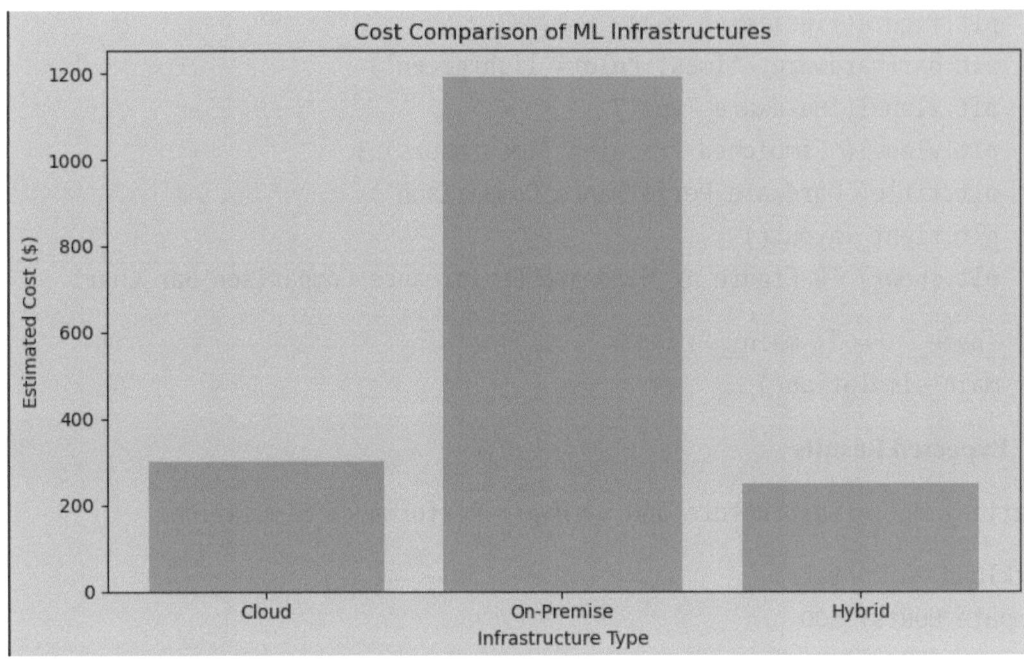

Figure 5-3. *Cost Comparison of ML Infrastructure*

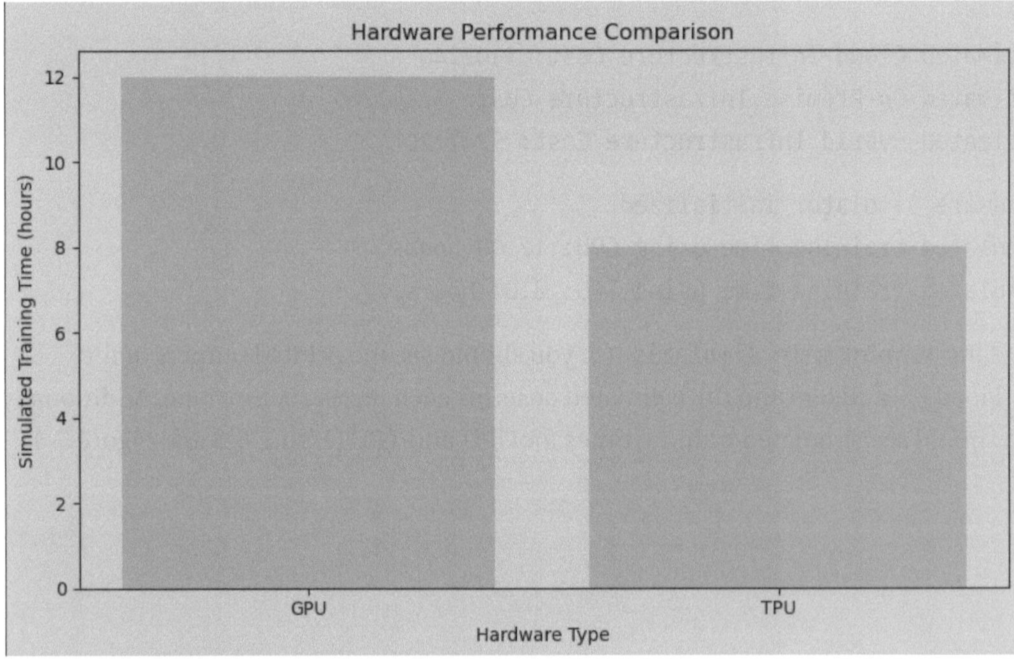

Figure 5-4. *Hardware Performance Comparison*

Conclusion

This project simulates ML infrastructure costs and hardware performance in a modular way. It helps you see how choices in workload, hardware, and infrastructure affect training time and total cost. For further exploration, you can use cloud provider calculators or extend the simulation to model compute, storage, and network usage over time to guide infrastructure decisions.

Summary

In this chapter, we explored ML infrastructure choices (such as cloud, on-premise, and hybrid), highlighting their trade-offs in scalability, flexibility, security, and cost. We discussed GPUs and TPUs, showing how hardware selection depends on workload, performance, and cost considerations.

We also covered cost optimization strategies, including efficient storage management, compute allocation, data transfer reduction, model pruning, quantization, and batch inference.

By applying these best practices, you can build scalable, cost-effective, and high-performance ML systems. The next chapter focuses on deployment strategies, including containerization, orchestration, and automation.

CHAPTER 6

Deployment Strategies for Machine Learning Models

Until now, we have discussed data pipeline, model selection, and the software and hardware parts of machine learning. Now we will dive deeper into strategies for machine learning model deployments. Here, we provide a deep investigation of deployment strategies for machine learning (ML) models by focusing on containerization, orchestration, and automation. We got familiar with these topics in Chapter 4, and here we dive deeper into them. Here, we start by reviewing the concept of containerization and tools such as Docker to package ML models for scalability and portability. Additionally, we discuss orchestration systems, such as Kubernetes. Kubernetes is examined for its capability in efficiently managing containers. We will provide some practical examples too, to demonstrate how to containerize and orchestrate ML models in practice. This chapter also discusses the design of deployment pipelines and highlights the role of continuous integration (CI) and continuous deployment (CD) tools for automating the deployment process. Practical samples, like automating the deployment of a fraud detection model, show these concepts in practice. In addition, techniques for maintaining and monitoring deployed models are provided. These techniques include tools for detecting and fixing data drift, retraining models, and managing versions. By addressing the most common challenges and providing actionable solutions, this chapter helps you learn efficient and robust deployment techniques for your ML systems.

Modern machine learning solutions not only need well-designed models but also require a robust system for packaging and deploying. This is where containerization **and** orchestration are getting important. Containerization and orchestration offer the

M. R. Mahdiani, *Mastering Machine Learning Architecture and Solutions*,
https://doi.org/10.1007/979-8-8688-2527-9_6

scalability, flexibility, and reliability that are needed for professional machine learning deployments. Here, we start with exploring how containerization technologies such as Docker and orchestration systems such as Kubernetes help in the deployment of machine learning models.

Containerization and Orchestration

Here, we review the concept of containerization and orchestration, and we make a deep and practical investigation.

Containerization: Packaging Models for Portability

I remember some years ago (maybe more than some years!) when I was a junior developer, using containerization was not as common as today. There, as a developer, I frequently had the problem that one code or software was working fine on my machine but not on my teammates' computers. You may have experienced similar issues. Containerization solves this problem.

Containerization is a method for deploying machine learning (ML) models. This method allows ML models to be packaged with all necessary libraries and dependencies into a single, portable unit. This approach guarantees that models run without any problem in different environments. Different environments can be on cloud platforms, local machines, or edge devices. By encapsulating the model, system libraries, runtime, and configurations, containerization gets rid of the "it does not work on my machine" problem.[29]

What Is Containerization?

Containerization is the technique of encapsulating an application, with all its dependencies, in a standard unit called the container. In contrast to traditional virtual machines, containers are not heavy, and they share the host OS kernel. They provide an isolated environment for running applications across different platforms. In machine learning, this means that the model that has been developed in a development environment could easily be packed and transferred to the production environment without worrying about its compatibility issues.

Docker for Machine Learning

One of the most common containerization platforms is Docker. By using Docker, data scientists and developers can create a Docker image that contains the trained model, the required dependencies (e.g., Python, scikit-learn, TensorFlow), and all environment-specific configurations.[30] The following code shows a typical Docker file that can be used to containerize a simple machine learning model.

Listing 6-1. Sample Docker File

```
# Base image
FROM python:3.9-slim

# Set the working directory
WORKDIR /app

# Copy the requirements file
COPY requirements.txt ./

# Install dependencies
RUN pip install --no-cache-dir -r requirements.txt

# Copy the rest of the code
COPY . .

# Expose the API port
EXPOSE 8080

# Run the application
CMD ["python", "app.py"]
```

In the above code, the Docker file starts with a Python 3.9-slim for the base image. You can use regular Python if interested. It installs the necessary dependencies specified in `requirements.txt`. In this file, you should provide the required dependencies (e.g., `scikit-learn==1.2.2`, `fastapi`, `uvicorn`). Finally, run the model inference API. By using this Docker file, you can build a Docker image with the following command. In this command, the -t flag is used to tag the Docker image with the name ml-model-api.

Listing 6-2. Common Commands for Docker

```
# Building and running a container
docker build -t ml-model-api

# Running the container on port 8080
docker run -p 8080:8080 ml-model-api
```

Expected Results

There are two commands in the above code, and each of them has its own expected results:

> **`docker build -t ml-model-api .`**: This command executes the instructions in the provided Docker file. Finally, it will return a message showing successful image building. You will see:
>
> `Successfully built [Image ID] and Successfully tagged ml-model-api:latest.`
>
> **`docker run -p 8080:8080 ml-model-api`**: This command starts the container. Output would be the logging of the app.py script. You can have access to the API endpoint at `http://localhost:8080` in your browser.

You can use containerization for different purposes, for example, for packing your code, transferring that to another machine, and training the model there. Now, we have a containerized model that can easily be deployed on any cloud platform or server.

If you have several containers, you can use Docker Compose which is very powerful and easy to use. Docker Compose lets you define several layers, such as server, database, and caching, in a single `docker-compose.yaml` file.

Listing 6-3. Sample docker-compose.yaml File

```
version: '3.8'

services:
  # Service for the machine learning API
  ml-api:
    build:
      context: .
      dockerfile: Dockerfile
```

```
    ports:
      - "8080:8080"
    depends_on:
      - db
    environment:
      DB_HOST: db
      DB_PORT: 5432
      DB_USER: postgres
      DB_PASSWORD: ****

  # Service for the PostgreSQL database
  db:
    image: postgres:13
    environment:
      POSTGRES_USER: postgres
      POSTGRES_PASSWORD: example
      POSTGRES_DB: ml_database
    volumes:
      - postgres_data:/var/lib/postgresql/data

volumes:
  postgres_data:
```

The above Docker Compose file defines and also configures two services and one volume for managing a simple application stack. The main service is ml-api that is the main machine learning API. This service looks for a Docker file in the current directory and maps port 8080 for accessing the API. It has a dependency service (db) which runs the database. The database in this example is PostgreSQL which uses a Docker volume Postgres data for persistently saving data.

Running and stopping Docker Compose is also very easy; you just need to run simple commands that are listed below.

Listing 6-4. Docker Compose Commands

```
# Running Docker Compose
docker-compose up

#Stopping Docker Compose
docker-compose down
```

Advantages of Containerization

Containerization has many advantages, especially if you are working on a big project. These benefits include, but are not limited to, scalability, portability, and isolation. Using containers, it would be easy to scale machine learning models horizontally. By horizontal scaling, I mean we add more computers to handle the load. We also have vertical scaling in which we give more resources (stronger CPU, more RAM, etc.) to a single computer. You just need to run several instances of the container. Containers can run on all platforms with installed Docker, which makes it easy to transfer models to production. Additionally, each container runs independently in an isolated environment. This guarantees that the dependencies of one model will not interfere with others.

Kubernetes for Orchestration

One of the most famous open source platforms for managing containers is Kubernetes. Kubernetes automates the deployment, operations, and scaling of application containers over clusters of hosts. While deploying machine learning models, Kubernetes guarantees that the suitable number of containers (replicas) are running, and the load is distributed efficiently over available resources.[31] The following code is a sample code of a Kubernetes configuration file (`deployment.yaml`) for deploying a machine learning model as a service. To run Kubernetes, you need a `deployment.yaml` file.

Listing 6-5. Sample `deployment.yaml` File

```
apiVersion: apps/v1
kind: Deployment
metadata:
  name: ml-model-deployment
spec:
  replicas: 3
  selector:
    matchLabels:
      app: ml-model
  template:
    metadata:
```

```
      labels:
        app: ml-model
    spec:
      containers:
      - name: ml-model-container
        image: ml-model-api:latest
        ports:
        - containerPort: 8080
```

The above configuration file defines a Kubernetes **deployment** that automatically manages three replicas of the containerized model (`ml-model-api`). Kubernetes efficiently handles load balancing in such a way that incoming requests are distributed efficiently across all running instances.

Key Features of Orchestration with Kubernetes

In this section, we talk more about orchestration with Kubernetes. We will tell you the reasons that you need to use Kubernetes for managing the containers.

In Chapter 4, we talked about microservices. Assume that you have ten different services, and each of them can have different loads. So you may need to assign more than one server to some of those services. Fortunately, using Docker, you can replicate the code in containers on different servers, but how can you manage them? What if some of them crash? What about the time that the load on one of them increases and on some other decreases? The response to all these questions is to use Kubernetes. Kubernetes automatically scales the number of containers based on demand and the number of requests in each of them. For example, if there is a growth in prediction requests, Kubernetes will add additional resources to handle the extra load. Additionally, Kubernetes offers built-in load balancing to route traffic to the appropriate containers based on availability. About service crashes, Kubernetes monitors and checks the health of each container and restarts them if they crash or fail.[32]

Now, let's assume a real-world example. Assume you have developed a machine learning model to predict customer churn. Docker is a good choice for containerizing the model and its dependencies, and Kubernetes is another good choice for deploying containers across multiple nodes. This manner helps your model handle high traffic. In addition, updating your system is easy, and you can monitor it efficiently. For the above system, you need to take the following steps:

1. **Develop and Train the Model**: Write code for developing your model, then train it using Jupyter Notebook or another tool.

2. **Containerize the Model**: Write a Docker file (similar to what we discussed earlier) and containerize your model.

3. **Deploy Using Kubernetes**: Create a Kubernetes structure to orchestrate your containers.

Figure 6-1 shows the integration of Docker and Kubernetes for deploying machine learning models. It illustrates the process from model development, containerization with Docker, to finally orchestrating using Kubernetes. This approach ensures scalability and reliability. This approach is very important and useful. Keep it in mind, and you will need it for your future projects.

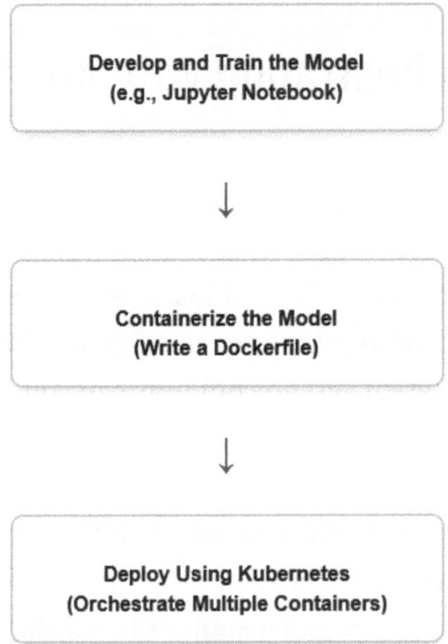

Figure 6-1. *Flowchart for Containerization and Orchestration in Machine Learning*

Deployment Pipelines and Automation

Deploying machine learning models efficiently needs a systematic strategy. This strategy should guarantee that models are correctly integrated into production environments and are capable of frequent updates with minimal risk. Deployment pipelines and automation play a crucial role in streamlining this process by offering an organized way for the continuous and consistent machine learning models. In this section, we will discuss how to create efficient deployment pipelines using automation tools. This approach enhances the efficiency and reliability of machine learning deployments.

Building a Deployment Pipeline

A deployment pipeline is a set of automated processes that get the code or machine learning models from the development environment and transfer them to production. Deployment pipelines are used to ensure that the model deployment is scalable, repeatable, and free from human error. The pipeline typically contains several main steps. In the following, we will talk about each step.

The first step is source control. First, machine learning models and related codes are committed to a version control system, like Git. This step allows tracking changes in different versions and ensuring that the committed model is up to date. Git is very useful, especially for a team. But I recommend using that in all projects, even if you are the only one who works on a project. It has happened several times that after a while, I remember a specific function that I had in a project before, and I do not have it anymore, but as I use Git, it is always easy for me to recover any part of my code.

The next step is continuous integration (CI). After the new code or model updates are committed, a CI server such as CircleCI, Jenkins, or GitHub Actions automatically builds the application and tests it. For machine learning projects, this includes running unit tests over model code, validating data transformations, and then testing inference against a validation dataset. If tests are successful, we go to the next step, which is model packaging. In this step, the model is packaged into a deployable unit. This step can involve containerizing the model using Docker, which was discussed earlier. Now the container is ready, and we can proceed with the next step which is continuous deployment (CD). In this step, the model is deployed to a staging environment or directly to the production environment. Tools like GitLab CI/CD, Jenkins, and ArgoCD help in automating the deployment process. The above procedure is important as it minimizes human intervention and reduces the risk of errors. Earlier, we talked about the churn prediction model. Remember it and assume the deployment pipeline for a customer churn prediction model. We can develop the pipeline using the following steps:

1. **Commit to Git Repository**: In the first step, the data scientist pushes their trained model to a Git repository.
2. **CI/CD Pipeline Triggered**: As the model is committed, a GitHub Actions or Jenkins pipeline is triggered. It builds, tests, and containerizes the model.
3. **Testing**: Now the model is built. So, the pipeline runs the validation tests to confirm if the model's accuracy is acceptable. This step could include comparing predictions on a validation set to ensure consistency.
4. **Deploy to Staging**: Then, if the tests are passed, the model will be deployed to a staging environment for integration testing. It is highly recommended to use a staging environment for more testing. Do not deploy your model directly to production.
5. **Deploy to Production**: At the end, as the final step, the model is deployed to the main production using tools such as AWS SageMaker or Kubernetes.

Like before, we will show the above steps in a figure for better visualization. Figure 6-2 shows the key steps in a machine learning deployment pipeline. In this figure, every step emphasizes automation to decrease human error and ensure consistent deployment across environments. It also shows the best practices and tools that enable transitions from development to production.

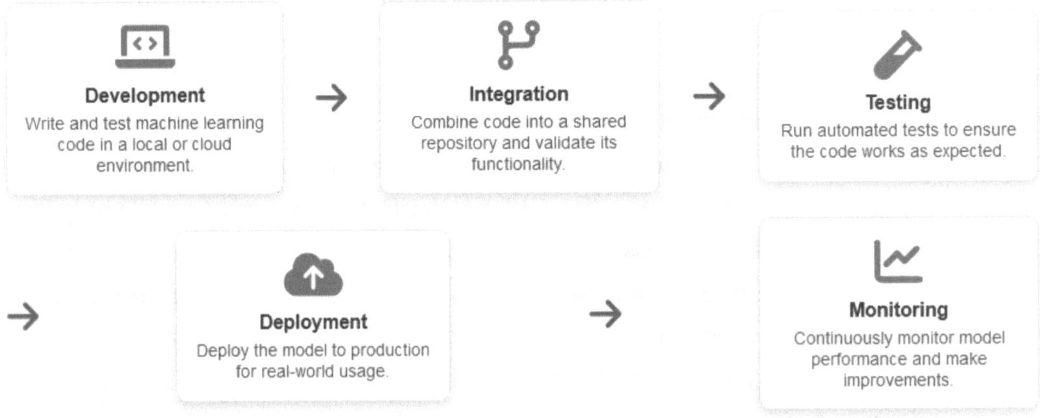

***Figure 6-2.** Deployment Pipeline Stages for Machine Learning Models*

Continuous Integration Tools

In the above, we mentioned CI/CD tools. We said that these tools can be used for automating integration and deployment. In this section, we discuss more about these tools. There are several tools for continuous integration. Jenkins and GitHub actions are two of the most famous ones. **Jenkins** is an open source automation server that can be configured for CI/CD workflows for machine learning projects. Jenkins pipelines can validate code, test models, and deploy them to various environments. Another tool is GitHub Actions, which is a native tool integrated into GitHub repositories on the GitHub website. The benefit of GitHub Actions is that it makes it simple to set up CI workflows directly within the GitHub version control system.

Continuous Deployment Tools

Also, there are some tools for continuous deployment that we will discuss, the most common ones here. Kubernetes, combined with Helm Charts, offers a strong tool for automating the deployment of machine learning models over clusters. Helm Charts make it simple to manage and version the Kubernetes resources that are needed for deploying models. Another useful tool for continuous deployment is ArgoCD. ArgoCD is a declarative continuous delivery tool that integrates perfectly with Kubernetes. ArgoCD offers a GitOps approach for managing application deployments. GitOps makes the continuous deployment of model changes easy. We talked about Jenkins before; now let's see how it works. Below is a simplified `Jenkinsfile` that shows how to automate the deployment of a machine learning model.

***Listing* 6-6.** Sample Jenkins File

```
pipeline {
    agent any

    stages {
        stage('Build') {
            steps {
                echo 'Building the Docker image...'
                sh 'docker build -t ml-model-api .'
            }
        }
```

```
        stage('Test') {
            steps {
                echo 'Running tests...'
                sh 'pytest tests/'
            }
        }

        stage('Deploy') {
            steps {
                echo 'Deploying to Kubernetes...'
                sh 'kubectl apply -f deployment.yaml'
            }
        }
    }
}
```

The above code does three main tasks. First, it creates a Docker image for the machine learning model. Afterward, it runs a set of tests to validate the model, and finally, it deploys the model to a Kubernetes cluster using a `deployment.yaml` file.

We discussed different steps for automating the model deployment, but why do we need automation? We will respond to this question in the proceeding.

Advantages of Automation in Model Deployment

In this section, we discuss the advantages of automation in model deployment. The first advantage is that the automation reduces manual errors. Automation guarantees that each deployment is run consistently, decreasing the chance of manual errors. Another advantage is faster time to market. Automated deployment pipelines allow quick testing and deployment of the models. It also decreases the time to production. And the third advantage is repeatability. Automation helps ensure that the deployment process can be repeated in various environments (development, testing, production) with minimal changes.

Let's talk about a real-world example. Consider a fraud detection model that needs frequent updates due to new fraud tactics. With automation, we can consider the following:

1. Scheduling the model retraining process every week.
2. Automatic test and validation of the model.
3. In the case of a validation pass, automatic model packaging for deploying to the production environment using Kubernetes. The CI/CD pipeline can take care of this part.

We talked about various tools that can help in deploying pipelines. Table 6-1 shows a summary of some tools that can be used in deploying pipelines.

Table 6-1. *Tools Commonly Used in Deployment Pipelines*

Tool	Function	Example Use Case
GitHub Actions	Continuous Integration	Triggering tests on code commits
Jenkins	Continuous Deployment	Deploying Docker containers
Kubernetes	Orchestration and Scaling	Running multiple replicas of a model
ArgoCD	GitOps-based Deployment	Automating production deployments

Now, we have deployed the model to production. Are we done? No, in the next section, we will talk about which task we should proceed after deploying the model to production.

Monitoring and Maintenance

After deploying the machine learning model to the production environment, there are other required tasks. Monitoring and maintenance are required to ensure that the model performs as expected and adapts to changing conditions. In this section, we explore various techniques and tools for maintenance and monitoring of the deployed machine learning models and preventing performance reduction over time.[33]

Importance of Monitoring Machine Learning Models

Why is there a need to monitor? Monitoring deployed models is required because the data and environment in production differ considerably from the conditions during model training. Changes in system performance, data distributions, and unexpected behaviors can affect the quality and accuracy of the model's predictions. So, monitoring helps in the early detection of these changes. Also, monitoring gives the data scientists the chance to take required actions before it affects the end users.

Types of Monitoring

There are several types of monitoring, and we discuss some of the most common ones here. The first type of monitoring is performance monitoring, which measures the performance of the model based on some predefined metrics like precision, recall, accuracy, and latency. A reduction in performance can show problems such as concept drift or data drift. Concept drift happens when the relationship between the features and labels changes over time. This usually needs fine-tuning or retraining the model to catch the new patterns. Sometimes, the data used in production diverges considerably from the training data. This is known as data drift. Tools like Evidently AI could be used to monitor and alert on data drift.

Maybe you ask, how do you detect concept drift and model drift? The answer is by monitoring the statistical properties of the input data. By comparing statistical properties of the features in production with those from the training dataset, you can indicate shifts in data patterns. This can be translated as data drift.

On the other hand, by tracking the distribution of model predictions and output performance, you can monitor concept drift.

Infrastructure monitoring monitors hardware and software metrics like memory usage, CPU, disk space, and network latency to verify that the underlying infrastructure is healthy and works properly. So these three items (data drift, concept drift, and infrastructure monitoring) are the most important items for monitoring. There are some other parameters, such as the number of user requests and details about them. Monitoring user requests is usually used for marketing.

Key Metrics to Monitor

To keep the model's performance in an acceptable range in production, it's required to track various metrics that show the health of both the model and the infrastructure. Here, we discuss some of the key metrics for monitoring.

Prediction Performance Metrics

Several parameters measure the performance; accuracy, precision, recall, and AUC-ROC are the most common ones.

> **Accuracy** measures the percentage of correct predictions. A decrease in accuracy could be an indicator of issues with the input data or a sign that there is a need for retraining.
>
> **Precision and Recall**: In classification models, precision and recall help in balancing between false positives and false negatives, respectively.
>
> **AUC-ROC**: The Area Under the Curve-Receiver Operating Characteristic (AUC-ROC) measures the model's capability to discriminate between classes. This is particularly helpful for evaluating binary classifiers.

Operational Metrics

There are some other parameters that monitoring them can be helpful; I call them operation metrics. These metrics include latency, throughput, and resource utilization. **Latency** measures the time required for the model to generate a prediction. High latency indicates performance bottlenecks. Throughput monitors the number of predictions in a unit of time. This indicates the model's efficiency under various loads. And finally, resource utilization means monitoring GPU, CPU, and memory usage to make sure that the system has enough resources to serve predictions.

Tools for Monitoring

Like other parts that we discussed earlier and introduced good tools, there are good tools for monitoring too. Some of the most common tools are Prometheus and Grafana. Prometheus is an open source alerting and monitoring tool, and Grafana offers a graphical interface for illustrating monitoring data. Together, these tools are commonly used for model performance and infrastructure monitoring. Another good tool for monitoring is Evidently AI. This tool is a monitoring tool that is specifically designed for tracking concept and data drift. It can integrate with different deployment platforms and provide insights into model stability. The third tool is AWS CloudWatch. For models deployed on AWS, CloudWatch can provide a real-time monitoring system for both application-level metrics and infrastructure. Similar to AWS CloudWatch, Azure Monitor offers comprehensive monitoring for Azure-based deployments. Azure Monitor collects data from applications and the underlying infrastructure and integrates well with machine learning models deployed on Azure ML.

The code below shows a sample Prometheus configuration file for setting up an alert system. This code will alert for the cases where the accuracy of the model falls below a specific threshold.

Listing 6-7. Prometheus Configuration File

```
- alert: ModelAccuracyDrop
  expr: model_accuracy < 0.85
  for: 5m
  labels:
    severity: critical
  annotations:
    summary: "Model accuracy is below the acceptable threshold"
    description: "The model accuracy has dropped below 85% for the last 5
    minutes. Immediate investigation is required."
```

In the above code, the alert `ModelAccuracyDrop` is triggered in the case that the model accuracy drops below 85% for five consecutive minutes. These alerts can be integrated with a notification tool such as Slack or email for prompt action.

Now, we have a system for monitoring the models; what is the next step?

Table 6-2 shows some of the most famous tools for monitoring.

Table 6-2. Monitoring Tools for Machine Learning

Tool	Purpose	Example Use Case	Key Features	Best Fit For
Prometheus	Infrastructure and Model Metrics	Monitor CPU usage and response time.	Time-series database; alerting capabilities	Monitoring distributed systems and applications.
Grafana	Visualization of Metrics	Plot model latency over time.	Customizable dashboards; integration with Prometheus	Visualizing complex metrics and trends.
Evidently AI	Data Drift and Concept Drift	Alert if feature distribution changes.	Automated drift detection; integrates with pipelines	ML model monitoring for feature and concept drift.
AWS CloudWatch	Cloud Infrastructure Monitoring	Track latency of deployed models.	Logs and metrics tracking; integration with AWS ecosystem	Monitoring cloud-native ML deployments.
Azure Monitor	Monitoring Azure Infrastructure	Analyze resource usage and ML model performance.	Integration with Azure services; query-based analysis	Teams using Azure cloud for ML deployments.
DataDog	Full-Stack Performance Monitoring	Detect anomalies in model prediction times.	Real-time monitoring; AI-driven anomaly detection	Comprehensive monitoring for ML and IT systems.
Neptune.ai	ML Experiment Tracking	Log and track model training experiments.	Collaboration tools; model versioning	Experiment-heavy ML research and development.
TensorBoard	Visualizing ML Training Metrics	Monitor loss and accuracy during model training.	Graph-based visualizations; scalability	Deep learning models built in TensorFlow.
Kibana	Visualization and Analysis of Logs	Analyze API call logs and request latencies.	Integration with Elasticsearch; real-time querying	Logging and operational monitoring.

(*continued*)

Table 6-2. (*continued*)

Tool	Purpose	Example Use Case	Key Features	Best Fit For
MLflow	ML Lifecycle Management	Track model performance across versions.	Experiment tracking; deployment management	End-to-end ML lifecycle workflows.
Alibi Detect	Outlier and Drift Detection	Identify adversarial attacks or anomalous data.	Pre-built detectors; explainability features	Robustness monitoring for production models.
Seldon Core	Deployment and Monitoring of Models	Monitor inference requests and responses.	Kubernetes-native; Model explainability	Scalable deployments on Kubernetes.

Best Practices in Monitoring

What should we do if we see something has gone wrong? We need to maintain the deployed model. In this part, we talk a little more about the best practices of monitoring and alerting. The most important point in monitoring is to define metrics that align well with the business goals. For instance, for a customer churn model (which we have some examples of before), precision and recall might be more important than accuracy. In Chapter 2, we talked about imbalanced data. Be careful that, in those cases, defining accuracy may not be suitable at all. Another good practice is to set up automated alerts for important metrics. Integrate alerts with communication tools for immediate notifications to the relevant stakeholders. Also, practice creating baseline metrics during the initial deployment phase. In there, you can define normal performance. This makes it more organized and easier to detect anomalies during the model execution in production. Finally, thresholds for alerts should be reviewed and updated regularly. This is required to reflect changing business goals or new data distributions. Static thresholds would become irrelevant over time as the model evolves and adapts.

Let's consider a more practical example. Consider a recommended system that is deployed for suggesting products to users on an ecommerce platform. To make sure that the model delivers relevant suggestions at all times, you should track prediction accuracy. Monitor various metrics such as click-through rate (CTR) as an indicator of the model's performance. Be careful about model drift. Use AI to detect when the distribution of the data of the user interaction changes. This indicates a need for model retraining. In addition, monitor resource utilization. Track memory and CPU usage with Prometheus to confirm that the model can handle the incoming traffic, particularly during peak shopping periods. And finally, set up alerts. Consider using Prometheus to configure alerts for sudden falls in high latency or CTR, and ensure timely intervention. Figure 6-3 shows the typical architecture for monitoring and alerting in a machine learning deployment. It represents different components involved in tracking model performance, identifying anomalies, and finally triggering alerts. Or in some cases, automating actions for maintaining operational integrity.

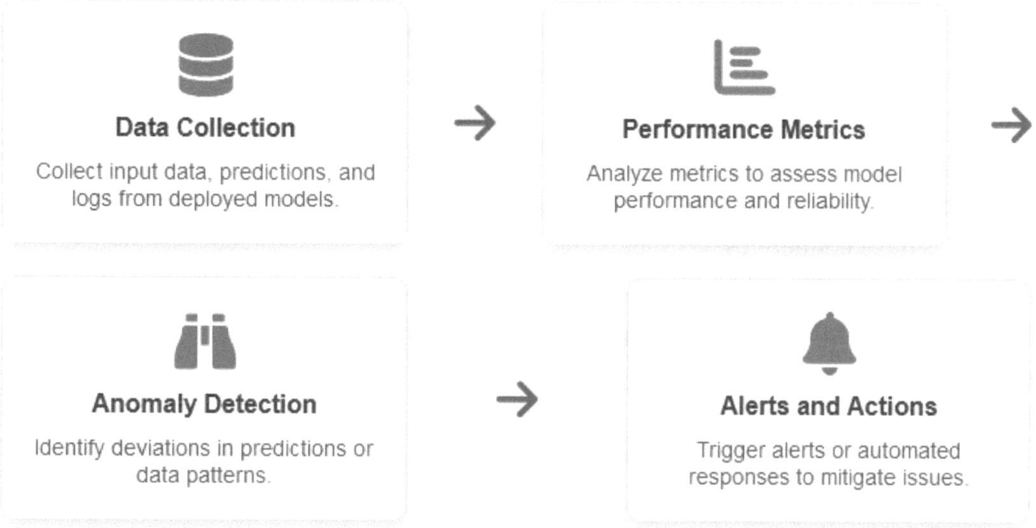

Figure 6-3. *Monitoring and Alerting Architecture for Machine Learning Models*

Challenges in Monitoring

Before finishing the monitoring part, we can talk about challenges in monitoring and alerting. One of the challenges is false alarms. In this case, wrongly set thresholds can cause frequent false alarms, which result in alert fatigue. It's crucial to tune alert thresholds carefully to avoid unnecessarily overwhelming the team. Another challenge is scalability. When the number of models grows, there is a need for the monitoring and alerting systems to scale accordingly. Using Kubernetes orchestration or cloud-based tools can help in managing scalability. Another challenge is data sensitivity. Tracking some metrics might need sensitive user data. Ensure that the monitoring and alerting systems follow the data privacy regulations like GDPR. One good practice for monitoring is to use human experts. This approach is very helpful in making robust models. In the next section, we talk about using human experts to monitor different parts of the ML workflow.

Human-in-the-Loop Machine Learning

Until now, we have had an extensive talk about monitoring. Here, we talk about a new approach for monitoring. Human-in-the-Loop (HITL) Machine Learning is a new paradigm that combines automated machine learning processes with human expertise and enhances model reliability and performance. HITL helps in active human involvement in the machine learning lifecycle. This approach is vital for applications that need high levels of human judgment, accuracy, or ethical consideration. By integrating the strengths of both machines and humans, HITL can result in more accurate and efficient models, particularly in complex scenarios where traditional machine learning techniques have challenges.

Human-in-the-Loop Machine Learning is achieved by integrating human feedback into the training, validation, and deployment phases in machine learning models. This strategy enables iterative improvements. In HITL, the machine learns from humans, and data offers additional information or corrections to improve the model. HITL improves the model's decision-making capabilities and reduces the risks associated with biased or incorrect predictions.[26]

The HITL strategy is especially effective in scenarios that involve subjective or ambiguous tasks that cannot be fully automated. For instance, in natural language processing (NLP) applications, humans can help resolve ambiguity in translating nuanced content or sentiment analysis. I will give an example here. Assume in a context, we want to know the sense of cool. It may be really cool or a terrible case that a person says is cool because of sarcasm. By involving human knowledge, the HITL system can learn to fix edge cases that would be problematic for an algorithm to handle independently. Now, think more and try to find other cases in which HITL can be a great help. In this part, we will discuss more situations in which HITL can be a great help.

Main Components of Human-in-the-Loop ML

Here, we discuss the main component of HITL. The first item is model training and feedback loop. In HITL, first, the model is trained using initial labeled data, then the human provides feedback to enhance the model's performance on misclassified or complex instances. This feedback loop proceeds until the model gains an acceptable level of accuracy. Another component is active learning. Active learning is a main aspect of HITL, in which the model selectively asks human experts for labels on data points that it finds uncertain or challenging. The third approach is iterative refinement. Human feedback is utilized iteratively to improve the model, making sure that it adapts to the changing conditions and learns from new data. This strategy is critical for applications that need continuous refinement.

Scenarios Where HITL Is Beneficial

Human-in-the-Loop Machine Learning is highly helpful in situations where human expertise is needed for decision-making or machine learning models cannot achieve acceptable levels of accuracy. Here, we represent some scenarios where HITL is advantageous.[34]

Handling Ambiguity and Subjectivity

In many machine learning tasks, there is inherent subjectivity or ambiguity. This makes it difficult for the model to predict accurately without human involvement. HITL provides clarity in these situations, which enables the model to learn the correct interpretation. For example, in sentiment analysis, text can contain ambiguity or sarcasm that the model fails to distinguish, similar to the example I mentioned a while ago about cool. Human annotators can correct these instances and provide valuable context that allows the model to learn more accurately.

Addressing Edge Cases and Rare Events

Machine learning models usually have a problem with rare events or edge cases that are not represented well in the training data. Human experts can offer important insights to help the model generalize well in these cases. For example, in self-driving cars, the model may see unexpected scenarios like unusual traffic patterns or rare road obstacles. Human intervention could help the model learn how to handle these rare situations safely.

Annotation and Labeling for Complex Datasets

HITL is also helpful for labeling and annotating complex datasets. Particularly, when the data involves nuanced interpretation or needs domain expertise. Human experts can guarantee the accuracy of labels, which is crucial for training high-quality models. For example, in medical imaging, radiologists can help label images to identify anomalies. This approach helps the model learn to detect these anomalies in new images more accurately. I prefer that for all data labeling human expert check and verify the labels.

Ethical Decision-Making and Bias Mitigation

In scenarios where ethical considerations are concerned, human intervention can increase a higher chance that the model's decisions are unbiased and fair. Be careful about the human experts that you select; humans can be biased, too. HITL can be used for identifying and addressing the biases in the model's predictions. This improves the ethical integrity. For example, in hiring models, humans can confirm the fairness of the model predictions to decrease the chance of biased predictions based on ethnicity, gender, or other factors.

Active Learning for Efficient Labeling

Active learning is a critical aspect of HITL that helps minimize the amount of labeled data needed for training. As we mentioned earlier, the model finds data points that are uncertain and asks human annotators for labels. So, it optimizes the labeling process. For example, in an image classification model, the model can be uncertain about some visually similar images. These images are then prioritized for human labeling, which will improve overall model performance with the least human involvement.

Human Involvement in Model Retraining

Model retraining is a crucial process for maintaining machine learning models relevant and up to date as new data becomes available. For example, assume that we have trained a model to help in trading; every day new trades happen and new data are available, and our model should be updated with this data. Human-in-the-Loop plays a crucial role in leading the retraining process to enhance model reliability and accuracy.[35]

Handling Data Drift and Concept Drift

We know that data drift and concept drift happen when the statistical parameters of the input data change over time. These drifts decrease the model's performance. HITL can help identify cases where the model's performance is degrading and provide corrective feedback. For example, in a credit scoring ML model, if the changes in economic conditions result in a change in customer behavior, human experts can help in detecting these shifts and offer feedback for retraining the model to make accurate predictions. We can have statistical parameters for detecting drifts too; however, for setting thresholds or involving the effect of some external parameters, such as political, we still need humans.

Incorporating Expert Feedback During Retraining

Human experts can give critical insights during model retraining, especially for complicated tasks in which model predictions require refinement. By involving human feedback, models could be retrained to address errors and enhance their overall performance. For example, in an image recognition system that is used for industrial quality control, human inspectors can check the model's predictions and find misclassifications. This feedback can be used for retraining the model and finally reducing false positives and false negatives. In all projects that I have been involved in, I consider expert consultancy as an indispensable part of designing machine learning. I suggest that if you oversee a serious ML project, from the beginning, start to talk to an expert in the field of your model. Experts usually provide very useful insights that are helpful in all steps of preparation, training, and monitoring your model.

Iterative Retraining for Continuous Improvement

HITL grants iterative retraining, where the model is continuously updated based on human feedback. This procedure helps ensure that the model remains accurate and relevant. For example, in an NLP app for customer support, human agents can check the model's responses to customer questions and correct them. These corrections are used in retraining the model, which improves its ability to handle similar queries in the future.

Figure 6-4 shows the Human-in-the-Loop (HITL) workflow. This figure emphasizes the iterative interaction between machine learning processes and human feedback. It shows how human intervention improves the model at different stages of the machine learning lifecycle, increasing reliability and accuracy.

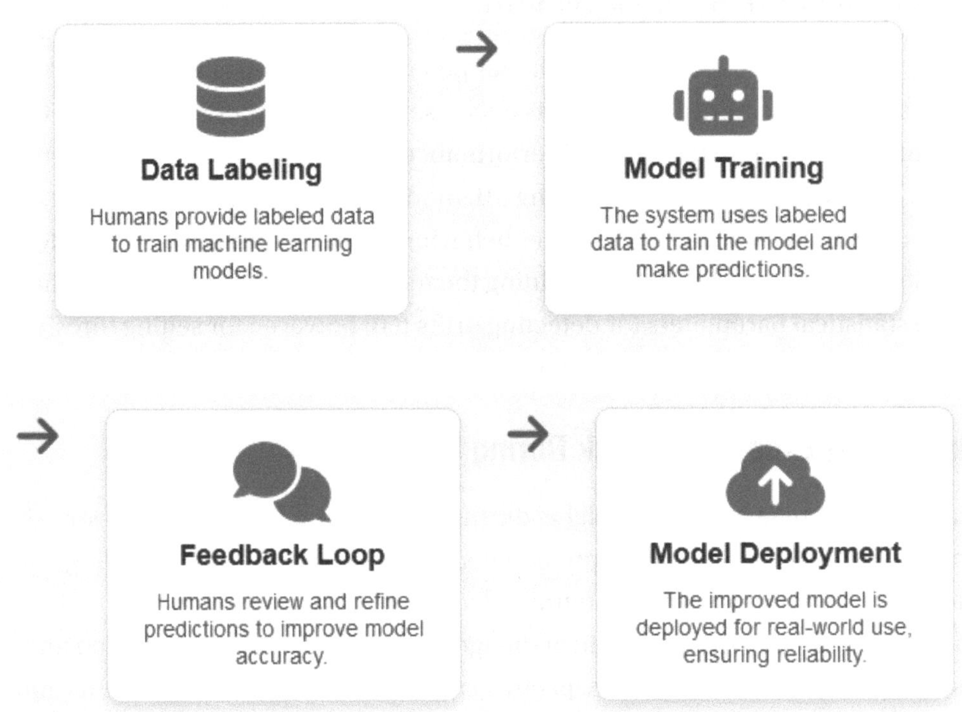

Figure 6-4. *Human-in-the-Loop Machine Learning Workflow*

Also, some scenarios where HITL is beneficial are listed in Table 6-3.

Table 6-3. *Scenarios Where HITL Is Beneficial*

Scenario	Description	Examples
Handling Ambiguity and Subjectivity	Addressing tasks with inherent ambiguity or subjective interpretation	Sentiment analysis, content moderation
Addressing Edge Cases and Rare Events	Providing insights into rare or unexpected situations	Autonomous driving, anomaly detection
Annotation for Complex Datasets	Ensuring accurate labeling of data requiring domain expertise	Medical imaging, legal document analysis
Ethical Decision-Making	Mitigating biases and ensuring fairness in model predictions	Hiring algorithms, credit scoring

Figure 6-5 shows the Human-in-the-Loop (HITL) workflow. It emphasizes the iterative interaction between machine learning processes and human feedback. It also shows how human intervention improves the model at different stages of the machine learning lifecycle.

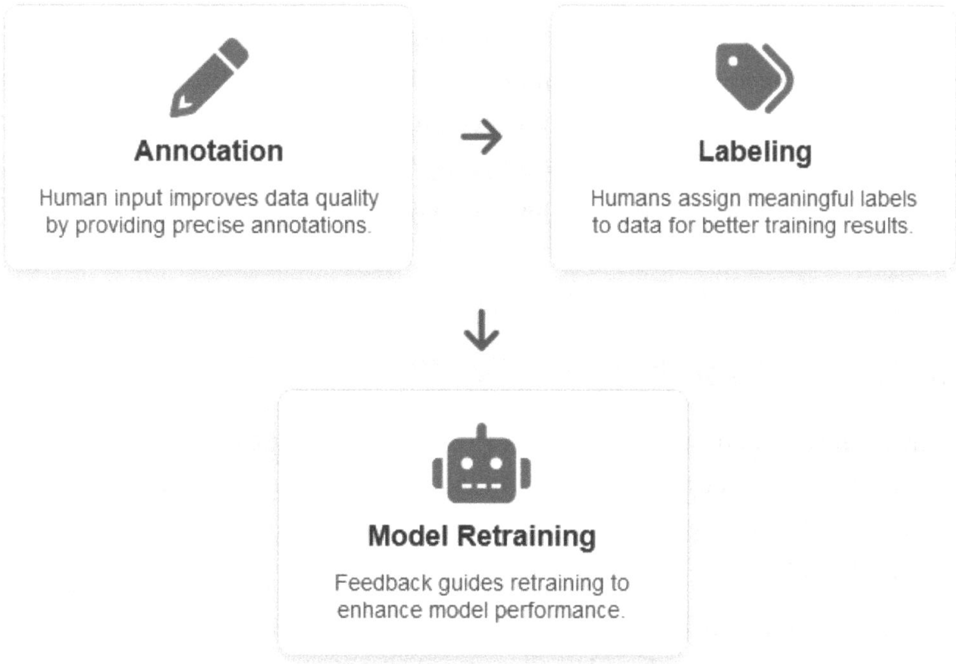

Figure 6-5. *Human-in-the-Loop Machine Learning Workflow*

Strategies for Collaboration in HITL

Collaboration between machine learning models and humans is required for maximizing the efficiency of Human-in-the-Loop Machine Learning. Good strategies for integrating model outputs with human feedback and designing user-friendly interfaces make a perfect interaction that improves the model's performance.

Combining Human Feedback with Model Outputs

Integrating model outputs with human feedback is an effective way to ensure reliability and enhance model performance. This collaborative approach helps the model to learn from human expertise, specifically in cases where the model is prone to errors or is uncertain.

Human experts can provide real-time feedback on model outputs, which allows immediate corrections. This is specifically beneficial in applications such as content moderation. In these types of applications, human moderators correct false positives or negatives that are generated by the model. But again, human feedback is beneficial in all types of ML projects.

Human feedback can be assigned various weights based on the expertise of the persons providing the feedback. For instance, the feedback from the domain experts can be given a higher priority compared to the feedback from a general user. This grants that the model learns from the most reliable sources. Do not forget that humans can make mistakes too. Weighted feedback is a good approach to minimize the effect of human mistakes. After talking about HITL benefits and best practices, we can go ahead with techniques to get human feedback effectively.

Designing User-Friendly Interfaces for HITL

Designing intuitive and user-friendly interfaces is critical for the best collaboration between machine learning models and humans. A well-designed interface helps human experts or annotators easily interact with the models, provide their feedback, and visualize model outputs.[36]

Annotation Tools: User-friendly annotation tools are required for effective feedback and labeling. Tools should help annotators with easy marking, categorizing, and providing comments on data points. Features such as zooming, drag-and-drop, and shortcuts can improve the user experience.

Visualization Dashboards: Some dashboards offer visual representations of performance metrics, model outputs, and areas of uncertainty. These dashboards make it easier for human experts to evaluate the model's behavior and provide targeted feedback. For instance, a dashboard showing model confidence scores helps experts focus on cases where the model is uncertain. However, be careful what you show to experts; showing some data may lead to biased feedback.

Feedback Interfaces: Interactive feedback interfaces, which allow users to make corrections directly to the model, are important for HITL workflows. These interfaces should be designed to reduce the required time and effort and help users provide feedback. This makes the process scalable and efficient.

Figure 6-6 shows the main strategies for Human-in-the-Loop (HITL) Machine Learning workflows. This figure emphasizes user-friendly interfaces and real-time feedback integration. These strategies improve the overall accuracy, efficiency, and usability of HITL systems.

Figure 6-6. *Collaboration Strategies in Human-in-the-Loop Machine Learning*

Challenges and Ethical Implications of HITL

Human-in-the-Loop (HITL) Machine Learning offers several benefits by integrating human expertise into automated systems; however, it also introduces some ethical implications and challenges. Solving these challenges is required for the effective, fair, and scalable HITL solutions.[37]

Addressing Bias Introduced by Human Intervention

While human intervention is valuable, as I mentioned before, it can introduce biases, too. These biases can affect the model's outcomes and predictions. Bias in the feedback process or in the data annotation can reinforce stereotypes or propagate discriminatory behavior. This leads to unfair model predictions.

Sources of Human Bias

Bias can arise from multiple sources, such as cognitive limitations, personal beliefs, and inconsistencies among human annotators. Since humans are not infallible, subjective judgments or errors can result in biased annotations that finally influence model training. For example, in a facial recognition system, human annotators might inadvertently label images unfairly based on gender or skin color. This leads to biases in the model. These biases can result in discriminatory behavior, particularly when the model is designed for sensitive applications such as law enforcement.

Strategies to Mitigate Human Bias

To mitigate bias, it is important to implement systematic approaches for fairness in the HITL process. Procedures to minimize human bias include

Diverse Annotation Teams: By employing diverse teams for data annotation, the biases of one group or individual do not dominate the whole training process. A balanced representation of perspectives helps create a more neutral dataset.

Annotation Guidelines and Training: By establishing clear annotation guidelines and training annotators based on those, you can decrease subjective interpretations and inconsistencies. This procedure helps keep the reliability and quality of labeled data.

Auditing and Bias Detection: Regular and professional audits of feedback and annotated datasets can help detect and correct potential biases. Also, some tools for bias detection can be integrated with the automatic flagging of problematic annotations.

Scalability and Efficiency Concerns

While HITL can considerably improve machine learning models, it also poses efficiency and scalability challenges. Human intervention needs considerable resources and time. This makes scaling HITL difficult for certain scenarios.

Limited Human Resources

Human annotators are a scarce resource, and as the volume of data increases, it gets challenging to handle the demand for human intervention. Large-scale datasets need extensive manual effort, which may result in bottlenecks in the machine learning workflow. For example, in industries such as autonomous driving, big datasets of driving footage require annotation. Scaling human annotation to cover all scenarios is practically impossible without introducing significant costs and delays.

Balancing Human and Machine Efficiency

Balancing automated processes with human intervention is crucial in maintaining efficiency in HITL workflows. Automation can help reduce the workload on human annotators. This goal can be achieved by reserving ambiguous or complex instances for human review while delegating routine tasks to machine learning models. More specifically, I am talking about active learning that we mentioned earlier. Active learning is a good practice for maintaining model quality while decreasing the workload.

Developing specific tools for feedback and annotation and integrating them into the HITL workflow speeds up the process. Keyboard shortcuts, efficient interfaces, and automated suggestions can decrease the time spent by annotators to provide feedback.

Figure 6-7 illustrates the main challenges of Human-in-the-Loop (HITL) systems. This figure shows bias and scalability concerns and, additionally, highlights strategies to overcome these challenges, like leveraging active learning techniques and employing diverse annotation teams for fairness and efficiency.

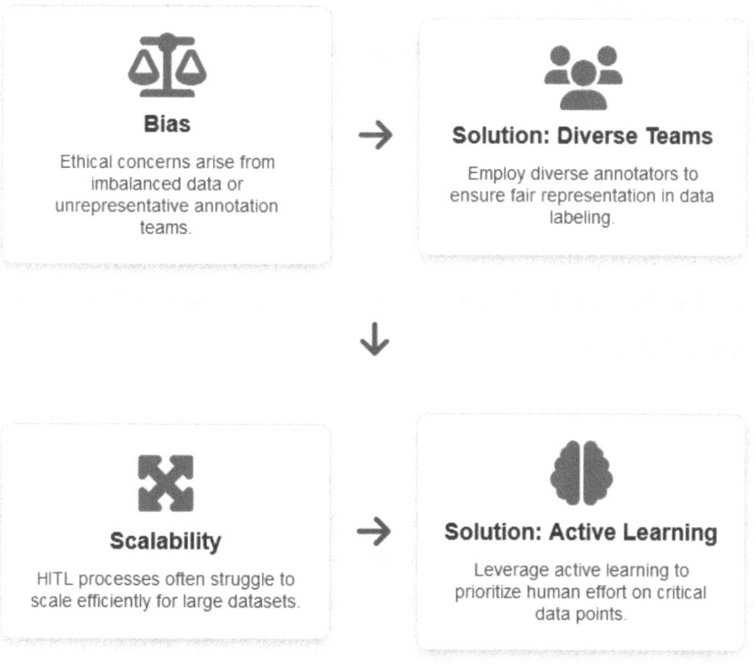

Figure 6-7. *Challenges in Human-in-the-Loop Machine Learning*

Maintaining Deployed Models

There is a need for machine learning models to be maintained periodically for long-term efficiency. Maintenance includes retraining models (when required), managing various versions, and updating pipelines. In the following, we discuss each maintenance topic in more detail. In this chapter, we just state the concept. In Chapter 7, we discuss maintaining deployed models in detail.

Retraining and Updating Models

Retraining models is required whenever data changes and subsequently performance metrics show degradation. Automated retraining can be scheduled, using cloud services or cron jobs, to keep the model up to date. Retraining frequency depends on the speed of the underlying data changes. A well-known approach is to set thresholds for the performance metrics—if the model performance is less than these thresholds, retraining should be triggered. Before we discussed monitoring and altering. We can set up alerting in a way that notifies us when we need retraining.

By retraining and improving models, we would have different versions of the model. Managing various versions of a model is crucial for facilitating A/B testing and backward compatibility. Model registries such as Amazon SageMaker Model Registry and MLflow provide a way to keep track of various versions of models. These tools make it easy to roll back to an earlier version if the current version underperforms or has an error or for any other reason we decide to go back to the previous version. We will have an extensive talk about versioning in Chapter 7.

Automation for Maintenance

Like all the parts discussed earlier, maintenance can be automated too. We can use continuous integration and continuous deployment (CI/CD) pipelines to automate model retraining and deployment processes. Try to automate as many processes as possible in your project. Automation keeps everything clean and saves tons of time and energy. For example, if a data drift is found, a new job for training, followed by testing and redeployment, can be triggered. For a better experience, keeping features up to date and consistent across training and production environments is crucial for model maintenance.

Here, we can discuss a simple real-world example for maintaining a machine learning model. Assume that a fraud detection model is deployed and its purpose is to identify fraudulent transactions. After deployment, the model's precision and recall should be monitored continuously to ensure it doesn't produce too many false negatives or positives. You can use Prometheus to track the latency and monitor the number of requests handled. Additionally, Evidently AI can alert whenever input data begins to shift, which means that there is a need for retraining. Additionally, alerts can be set up for the continuous monitoring of precision and recall metrics. In the cases that precision falls below a predefined threshold, alerts will notify the data scientist. The third monitoring item is for data drift detection. For drift detection, we need to compare the distributions of incoming data with the training dataset. So, whenever the distribution mismatch drift is detected, the system provides alerts. Figure 6-8 shows the key steps in a machine learning deployment pipeline. Each step uses automation to decrease human error and grant consistent deployment across environments. It emphasizes the tools and practices that enable smooth transfer from development to production.

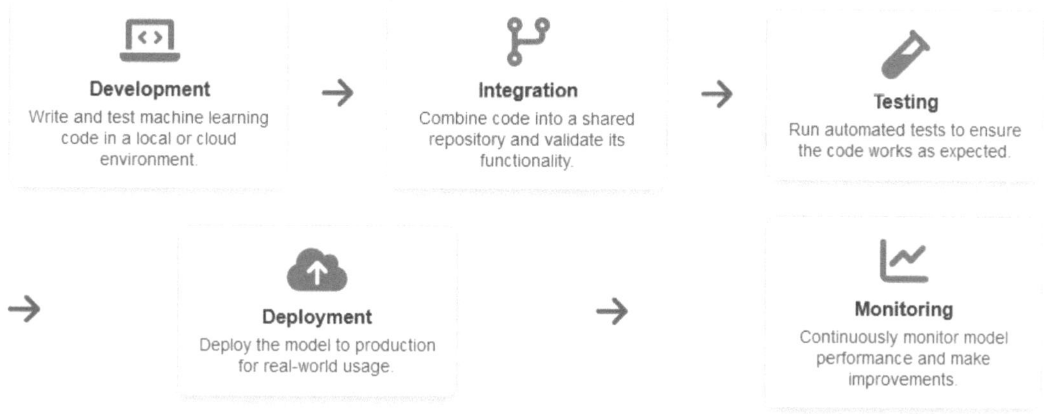

Figure 6-8. *Deployment Pipeline Stages for Machine Learning Models*

In this chapter, we talked about model deployment, monitoring, and automating the whole process. Like all other chapters, we finish it with a practical project.

Project Title: Automated Deployment and Monitoring of a Fraud Detection Model

Like the previous chapter, the project of this chapter is a simulation. Here, we will show a complete end-to-end deployment pipeline for a machine learning model.

Objective

Here, we focus on containerization, orchestration, deployment automation, and monitoring. We discussed all these topics in this chapter. In this project, you will test your learning in simulating the packaging of a model into a container, automating its deployment (mimicking CI/CD processes), and setting up monitoring for maintenance and performance.

Project Description

The project is divided into several steps that we discuss here.

Step 1: Import Required Libraries

We import necessary libraries for numerical computations, data simulation, and plotting. The code below imports the libraries. In this step, we imported `NumPy` for data operations and `matplotlib` for plotting simulated results.

Listing 6-8. Step 1: Import Required Libraries

```
import numpy as np
import matplotlib.pyplot as plt
import time
```

Expected Results

This step should run without producing any output; it just imports the libraries, so no result is a good sign that the code has not encountered any error.

Step 2: Simulate a Fraud Detection Model

In this step, Step 2, we simulate a simple fraud detection model. The model will be represented by a dummy class that can "predict" fraud based on a random threshold. To practice modular architecture, we define classes. The code below defines a class for simulating fraud detection model. In this code, we define the `FraudDetectionModel` class. This class simulates a fraud detection model. It has a constructor that simulates the training process and a `predict` method that generates random binary predictions. This simple model is a placeholder for a real fraud detection algorithm. We are not going to talk about fraud detection here; the purpose of this project is to show automated deployment and monitoring.

Listing 6-9. Step 2: Simulate a Fraud Detection Model

```
class FraudDetectionModel:
    def __init__(self):
        # Simulate model training time.
        print("Training Fraud Detection Model...")
        time.sleep(1)  # Simulate time delay for training.
        self.threshold = 0.5  # A dummy threshold for classification.
        print("Model training complete.\n")

    def predict(self, data):
        """
        Simulate predictions on input data.

        Args:
            data (numpy.ndarray): Input features.

        Returns:
            numpy.ndarray: Binary predictions indicating fraud (1) or non-
            fraud (0).
Model Versioning        """
        # For demonstration, generate random predictions based on a fixed
        threshold.
        predictions = (np.random.rand(len(data)) > self.threshold).
        astype(int)
        return predictions
```

```
# Instantiate the model
fraud_model = FraudDetectionModel()
```

Expected Results

```
Training Fraud Detection Model...
Model training complete.
```

After running this step, you should see the output that indicates the fraud detection model is being trained. It also should show another message that confirms the training is complete.

Step 3: Simulate Containerization of the Model

The code below defines a class for simulating containerization. This code defines the ModelContainer class for simulating the containerization process. This class gets the model and a container name. The run method simulates the execution of the container by calling the model's prediction method. This is how models are containerized in practice.

Listing 6-10. Step 3: Simulate Containerization of the Model

```
class ModelContainer:
    def __init__(self, model, container_name="fraud_detection_container"):
        """
        Initialize the container with the model.

        Args:
            model: The ML model to be containerized.
            container_name (str): A name for the container.
        """
        self.model = model
        self.container_name = container_name
        print(f"Container '{self.container_name}' created for the
        model.\n")

    def run(self, input_data):
        """
        Simulate running the container to make predictions.
```

```
        Args:
            input_data (numpy.ndarray): Input features for prediction.

        Returns:
            numpy.ndarray: Model predictions.
        """
        print(f"Container '{self.container_name}' is running the model...")
        predictions = self.model.predict(input_data)
        print("Container execution complete.\n")
        return predictions

# Wrap the fraud detection model in a container
model_container = ModelContainer(fraud_model)
```

Expected Results

```
Container 'fraud_detection_container' created for the model.
```

When you run the code of this step, the console output should show that the container is created and is ready to run predictions.

Step 4: Build a Deployment Pipeline Simulation

The code below defines a class for simulating the deployment pipeline. In this code, the `DeploymentPipeline` class simulates a CI/CD pipeline to deploy our containerized model. It has several methods:

- `build`: Simulates the process of building a container image
- `test`: Simulates container tests
- `deploy`: Simulates deployment to production
- `run_pipeline`: Combines all other steps for deploying the container and running the model

Listing 6-11. Step 3: Simulate Containerization of the Model

```
class DeploymentPipeline:
    def __init__(self, container):
        """
        Initialize the deployment pipeline with the containerized model.

        Args:
            container (ModelContainer): The container holding the model.
        """
        self.container = container

    def build(self):
        print("Building the container image...")
        time.sleep(1)  # Simulate build time delay.
        print("Container image build complete.\n")

    def test(self):
        print("Running tests on the container...")
        time.sleep(1)  # Simulate testing time delay.
        # For demonstration, assume tests always pass.
        print("All container tests passed.\n")

    def deploy(self):
        print("Deploying the container to production...")
        time.sleep(1)  # Simulate deployment time delay.
        print("Container deployed successfully.\n")

    def run_pipeline(self, input_data):
        """
        Run the entire deployment pipeline and execute the container.

        Args:
            input_data (numpy.ndarray): Data for the model to process.

        Returns:
            numpy.ndarray: Predictions from the deployed container.
        """
        self.build()
```

```
        self.test()
        self.deploy()
        # Run the container to get predictions.
        predictions = self.container.run(input_data)
        return predictions

# Create a deployment pipeline instance using the containerized model
pipeline = DeploymentPipeline(model_container)

# Simulate some input data for prediction (e.g., 10 samples with dummy
features)
input_data = np.random.rand(10, 4)
```

Expected Results

Running the pipeline's methods must print messages for each step of the process: building, testing, deploying, and finally running the container to obtain predictions.

Step 5: Simulate Monitoring and Maintenance

The code below defines a class for simulating monitoring and maintenance. In this code, the `MonitoringSystem` class simulates the monitoring of the deployed model. It checks the metrics such as error and latency over several iterations. The `record_metrics` method logs all metrics, and `display_metrics` shows them on plots. This simulation helps understand how monitoring systems track model health and performance over time and alert to any anomalies.

Listing 6-12. Step 3: Simulate Containerization of the Model

```
class MonitoringSystem:
    def __init__(self):
        # Initialize an empty list to store metric logs.
        self.metrics_log = []
        print("Monitoring system initialized.\n")

    def record_metrics(self, latency, error_rate):
        """
        Record performance metrics.
```

```
        Args:
            latency (float): Simulated response time of the deployed model.
            error_rate (float): Simulated error rate of model predictions.
        """
        metric = {"latency": latency, "error_rate": error_rate}
        self.metrics_log.append(metric)
        print(f"Recorded metrics: Latency = {latency} sec, Error Rate =
        {error_rate}%")

    def display_metrics(self):
        # Plot the recorded metrics for visualization.
        if not self.metrics_log:
            print("No metrics recorded.")
            return

        latencies = [m["latency"] for m in self.metrics_log]
        error_rates = [m["error_rate"] for m in self.metrics_log]
        iterations = range(1, len(self.metrics_log) + 1)

        # Plot latency
        plt.figure(figsize=(8, 4))
        plt.plot(iterations, latencies, marker='o', linestyle='-',
        color='blue')
        plt.xlabel("Iteration")
        plt.ylabel("Latency (sec)")
        plt.title("Model Response Latency Over Time")
        plt.tight_layout()
        plt.show()  # Figure 4: Latency Monitoring Plot

        # Plot error rate
        plt.figure(figsize=(8, 4))
        plt.plot(iterations, error_rates, marker='o', linestyle='-',
        color='red')
        plt.xlabel("Iteration")
        plt.ylabel("Error Rate (%)")
        plt.title("Model Prediction Error Rate Over Time")
```

```
        plt.tight_layout()
        plt.show()  # Figure 5: Error Rate Monitoring Plot

# Instantiate the monitoring system
monitoring_system = MonitoringSystem()

# Simulate periodic monitoring by recording metrics
# For the purpose of demonstration, we record metrics over 5 iterations.
for i in range(5):
    simulated_latency = np.random.uniform(0.2, 1.0)  # simulated latency
    between 0.2 and 1.0 seconds
    simulated_error_rate = np.random.uniform(0, 5)     # simulated error
    rate between 0% and 5%
    monitoring_system.record_metrics(simulated_latency, simulated_
    error_rate)
    time.sleep(0.5)  # Simulate time gap between recordings
```

Expected Results

```
Monitoring system initialized.

Recorded metrics: Latency = 0.8377063260983804 sec, Error Rate =
0.8795552324930628%
Recorded metrics: Latency = 0.3527441087668747 sec, Error Rate =
3.014518091488145%
Recorded metrics: Latency = 0.7209030147677165 sec, Error Rate =
0.9055356439254231%
Recorded metrics: Latency = 0.6829959220938646 sec, Error Rate =
2.5661426730412886%
Recorded metrics: Latency = 0.22399632773473455 sec, Error Rate =
4.1166214440486595%
```

This step has output, and you should see the printed messages for each metric that is recorded during the simulation. After that, there are two plots: one shows latency over time (Figure 6-9), and the other shows the error rate (Figure 6-10).

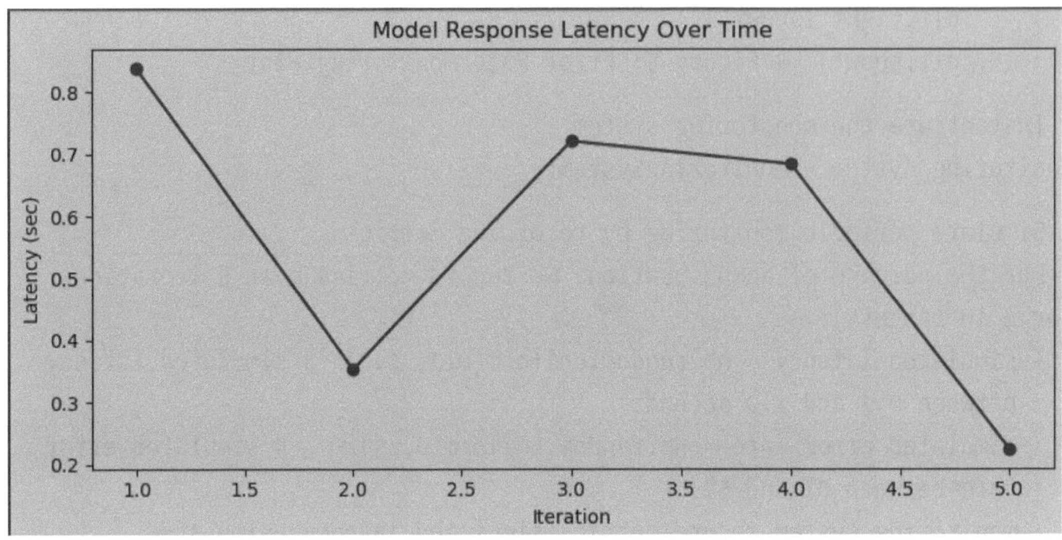

Figure 6-9. *Model Response Latency Over Time*

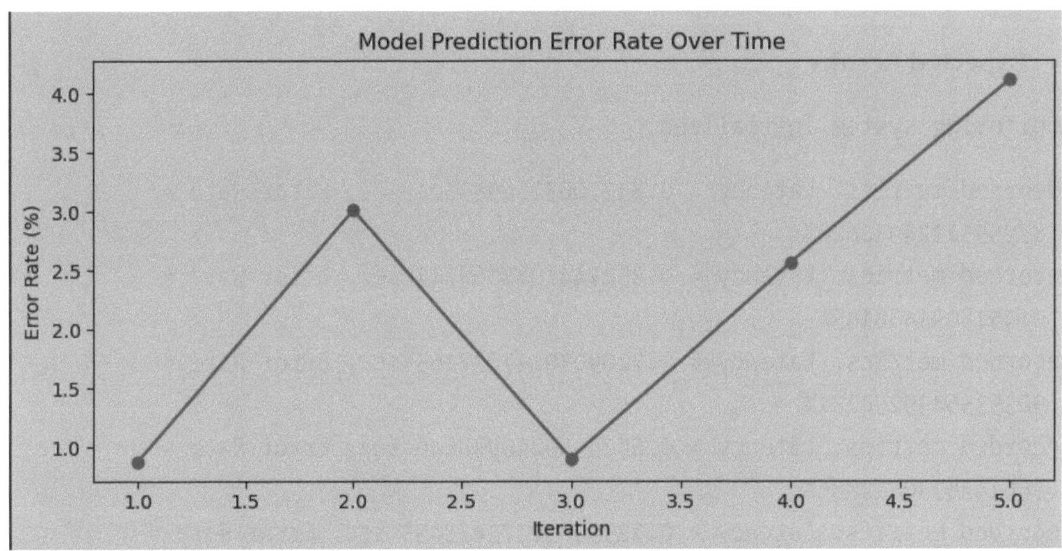

Figure 6-10. *Model Prediction Error Rate Over Time*

Step 6: Orchestrate the Entire Deployment and Monitoring Pipeline

The code below orchestrates the entire deployment and monitoring pipeline. In this code, the `main_deployment()` function orchestrates the whole process. It runs the deployment pipeline, which includes container building, testing, and deploying, and then runs the model to generate predictions. Afterward, it calls the monitoring system to display the performance metrics which are collected earlier. This step simulates the full deployment scenario, including packaging, deploying, and monitoring in production.

Listing 6-13. Step 3: Simulate Containerization of the Model

```
def main_deployment():
    print("=== Starting the Automated Deployment and Monitoring
    Simulation ===\n")

    # Run the deployment pipeline to deploy the containerized model and
    obtain predictions.
    predictions = pipeline.run_pipeline(input_data)
    print("Predictions from the deployed model:")
    print(predictions, "\n")

    # After deployment, simulate monitoring of the deployed model.
    print("Simulating monitoring of the deployed model...\n")
    # Display the recorded monitoring metrics.
    monitoring_system.display_metrics()

    print("=== Simulation Complete ===")

# Execute the main deployment simulation
if __name__ == "__main__":
    main_deployment()
```

Expected Results

```
=== Starting the Automated Deployment and Monitoring Simulation ===

Building the container image...
Container image build complete.

Running tests on the container...
All container tests passed.

Deploying the container to production...
Container deployed successfully.
```

```
Container 'fraud_detection_container' is running the model...
Container execution complete.

Predictions from the deployed model:
[0 0 0 0 0 0 0 0 0 1]

Simulating monitoring of the deployed model...

=== Simulation Complete ===
```

After you run `main_deployment()`, you will see the messages from the code for each step. You can also see the printed predictions that are made by the deployed fraud detection model and plots for displaying monitoring metrics.

Conclusion

This project shows a complete simulation of deploying a fraud detection model. It uses modern deployment techniques. We simulated all containerization, an automated CI/CD-style deployment pipeline, and a monitoring system. Each component is modular and is designed for understanding the main concepts of the chapter, such as containerization, orchestration with CI/CD pipelines, and model maintenance and monitoring.

By following the steps of the project, you can test your knowledge and apply these methods in real-world ML deployment scenarios. You can make it more challenging by using the tools that we discussed in the chapter. In this way, you will deepen your knowledge.

Summary

This chapter has emphasized the importance of packaging, maintenance, and monitoring in machine learning deployments. We talked about how to make a container, what is the importance of containers, and how to orchestrate them. Using these techniques, we practiced a modular ML system. Afterward, we talked about tools and techniques to automate the process. We discussed what continuous integration is and how continuous deployment can be implemented. We also investigated how CI/CD can be effective in increasing the performance of an ML system. In addition, we discussed effective monitoring and automating all processes. This allows data scientists to actively detect and address issues like concept and data drift. Additionally, maintenance strategies such as automated retraining and versioning help models remain reliable and accurate. Tools such as Prometheus and Grafana, and Evidently AI, in addition to CI/CD automation, provide robust support for keeping models effective and accurate over time. We also talk about Human-in-the-Loop in monitoring and its effect on making robust models.

In the next chapter, we will talk about the best practices for deploying and maintaining a model in production.

CHAPTER 7

Managing and Updating ML Models in Production

From the beginning of the book until this chapter, we proceeded step by step and covered many important topics. In Chapter 6, we talked deeply about automating machine learning (ML) pipelines. In this chapter, we will discuss the best practices and strategies for updating and managing ML models in production environments. We start with an investigation of model versioning, which is essential for traceability and consistency. Then, we discuss best practices, including metadata storage, semantic versioning, and use of model registries, with tools like DVC, MLflow, and Kubeflow. Real-world examples show how to implement versioning efficiently.

This chapter also talks about the critical necessity for retraining models and investigates why and when retraining is necessary. We mentioned earlier about cases like data drift and performance degradation. We will talk about techniques for retraining, such as full retraining, incremental learning, and transfer learning, and we will investigate some examples and workflows. Additionally, we talk about automation tools, such as Airflow and MLflow, which are used to streamline retraining and updating processes. Finally, we talk about how to examine challenges and provide actionable insights. In this chapter, you will gain the required knowledge about maintaining and enhancing the performance of ML models in production environments. Firstly, we start with model versioning.

Model Versioning

We had a quick talk about model versioning in Chapter 6; here, we go into more details. As you know, deploying a machine learning model to production is not the end. Versioning is important because after deploying the model, new techniques are

M. R. Mahdiani, *Mastering Machine Learning Architecture and Solutions*,
https://doi.org/10.1007/979-8-8688-2527-9_7

developed, data changes, and models require improvements. Model versioning helps data scientists track, manage, and control several iterations of a model. In this part, we will talk about the significance of versioning, best practices, and available tools for streamlining this process.

Importance of Model Versioning

There are several reasons for the importance of model versioning. Versioning helps track improvements and changes to a model over time. They make all modifications documented. With good versioning, it would be easy for us to understand why and how each version of a model is created and deployed.[38] In addition, versioning helps in comparison and experimentation. Versioning models allows for different iterations to be tested, and we can compare them with each other. For example, a new model could be deployed with the existing one to compare their performance. This practice is known as A/B testing. One other reason for versioning is rollback capability. If a new version of a model does not work as expected, versioning makes it possible to roll back to a previous stable version, which is very important in production systems.

Versioning Best Practices

After reviewing the importance of versioning, it is time to discuss how to manage model versioning effectively. There are several best practices for this purpose.

Use Semantic Versioning

Semantic versioning is widely used in software development and can be used for models, too. Usually, a semantic version is in the following format: MAJOR.MINOR. PATCH. MAJOR shows a significant change, such as a new model architecture. MINOR shows smaller changes or improvements, such as retraining with the bigger dataset or changing hyperparameters. And finally, PATCH shows minor changes or bug fixes, like fixing a feature scaling problem. For example, a model that changes from version 1.0.0 to 1.1.0 has minor improvements; however, a move from 1.0.0 to 2.0.0 undergoes significant updates. In the case of just fixing a bug, the version can change from 1.0.0 to 1.0.1.

Store Metadata for Each Version

Another best practice in versioning you should follow is storing metadata. These metadata include training data information, model parameters and hyperparameters, evaluation metrics, and environment details.

Training data information includes detailed information about the dataset, such as preprocessing steps and date ranges. Model parameters and hyperparameters are settings specifically used for training, like batch sizes and learning rates. Evaluation metrics include precision, recall, and accuracy. These metrics are necessary for comparison between different versions. Furthermore, environment details, including information about the tools and libraries used during training, help to replicate the training environment in the future.

We mentioned model registry before, but what is it, and why is it needed? We will talk about it in the next section.

Use a Model Registry

The model registry is a centralized repository for storing and managing different versions of our machine learning models. Its role is to keep records of model iterations, such as deployment status and metadata. Some of the most common model registries are MLflow and Amazon SageMaker. I highly recommend using a model registry in projects, especially if your project is a serious, big project. If you are using the AWS ecosystem, SageMaker is a good choice; if not, MLflow can be suitable. MLflow model registry offers a model registry to maintain versions, track models, and manage lifecycle transitions (e.g., from staging to production). Amazon SageMaker Model Registry helps data scientists register, evaluate, and track versions of models in AWS SageMaker. Using a model registry helps standardize the model deployment process and enables more efficient collaboration with other data scientists or engineering teams. Anyway, there are many benefits of model registries, and they play the role of Git for machine learning models. In the next part, we discuss how to use these tools.

Tools for Managing Model Versions

Earlier, we discussed model versioning tools such as MLflow and SageMaker model registry. Here, we talk in detail about how to use model versioning tools.

MLflow

MLflow is an open source platform that helps track experiments and package the code for reproducible runs. It also manages models with a model registry. Using MLflow, you can compare various versions of models and easily manage transitions, like transferring a model from development to production.[2,39]

DVC (Data Version Control)

Another tool is DVC, which integrates with Git for managing various versions of datasets, machine learning models, and experiments. DVC allows maintaining a link between the data, the code, and the model. DVC makes reproducing experiments and rolling back to previous versions easy.

Kubeflow

Another good tool is Kubeflow. Kubeflow is a cloud-native platform that can be used for managing, deploying, and scaling machine learning models on Kubernetes. Its Kubeflow Pipeline (KFP) is a powerful tool in tracking the entire linage including data, code, and steps which is crucial in versioning and necessary for reproducing results.

DVC has strong versioning capabilities, which facilitate tracking various versions of models and deployments. The following code shows how to register a model using MLflow.

Listing 7-1. Registering a Model in MLflow

```
import mlflow
import mlflow.pyfunc

# Load the trained model
model = mlflow.pyfunc.load_model("models:/my_model/Production")

# Register a new version of the model
mlflow.register_model(
    "runs:/<run-id>/my_model",
    "MyModelRegistry"
)
```

The above code registers a new version of the model in the MLflow model registry. This action makes the model available for tracking and deployment. Now we can talk about a practical example of using versioning. Assume that you have a machine learning model that predicts the price of houses. Over time, you have more data, and you add new features, so you need to create new versions of your model. You can practice versioning as follows. First, for the initial deployment, assign version `1.0.0` to the model. We assume this model is trained using historical data from 2019 to 2021, and it is also deployed to production. For your information, in some cases, we use other versions such as 0.0.1. The versions that start with 0 usually change a lot and are not deployed to production. After a while, you update the data, and you release a minor version (`1.1.0`). For this case, we assume that the model is retrained with data from 2019 to 2022 and has a slight improvement in accuracy. Afterward, you may add a new feature and release a major version (`2.0.0`). These new features can be, for example, adding economic indicators to the model, which leads to significant performance improvement. And finally, you fix a bug and release a patch version (`2.0.1`) to fix a small issue related to the feature scaling. I hope with this example you can understand the versioning better. Versioning can have some challenges, too, which we discuss in the next section.

Challenges in Model Versioning

One of the challenges in versioning is storage management. Each version of a model has a volume and needs storage, which can be expensive when dealing with big models or a huge number of versions. To overcome this issue, you can use cloud storage solutions, and you can prune outdated versions. Another challenge is version sprawl. Version sprawl is the situation where the number of versions rapidly grows and changes in which tracking is more difficult. Without semantic versioning techniques, maintaining track of several versions can become difficult. Use a model registry and use a strict versioning policy to mitigate this issue. The next challenge is collaboration across teams. In cases that multiple teams are involved, versioning becomes more complicated. Clean documentation and a centralized version tracking system are required to help everyone work with the correct version of the model. So now, you know how to version ML models and data, what the importance of versioning is, and what the challenges are. In addition, you know how to overcome challenges. You know, one of the main reasons for having several versions is that we retrain the models. In the next section, we will talk more about retraining.

Retraining and Updating

After a machine learning model is deployed, it is critical to be sure that the model predictions remain effective and accurate over time. Updating and retraining the models in production is the main factor for keeping their reliability, specifically in dynamic environments. In dynamic environments, data distributions can fluctuate, and new information with new patterns may arrive. In this section, we discuss why retraining is required, when to retrain models, and tools and strategies for retraining.

Why Retraining Is Necessary

Usually, machine learning models are trained over historical or a specific dataset. Thus, the performance of these models is good only in cases where the data that is sent for inference is similar to the data that is used for training. But in real-world scenarios, data evolves, which leads to performance degradation.[35] One reason that makes retraining necessary is data drift. We have mentioned data drift before. In data drift, the characteristics of the input data change, which results in the model underperforming. For instance, assume a model is trained to predict customer behavior. This model needs retraining when consumer preferences change over time. Another case is concept drift. In concept drift, the relationship between features and the labels changes. For example, in a fraud detection model, the model may require retraining if fraudsters find and use a new detection technique. Another case is the availability of the new data. These data can be provided from various resources such as a logging system or sensors. After manually cleaning the data, they can be used for retraining the model. If new data is available and can improve the model's accuracy, retraining with this updated data can be useful to ensure that the model works based on the latest data. Also, when we have performance degradation, we may need model retraining. Monitoring metrics like precision, recall, and accuracy can help identify a decline in model performance. In these cases, retraining is required to restore the model's efficacy. After investigating different reasons and situations that require retraining, we can talk about timing.

When to Retrain Models

To select the best time to retrain a model, you should consider the application and the specific needs of the system. In the following, we discuss the common strategies for finding out when to retrain your model.

Scheduled Retraining

In environments in which data changes predictably and slowly, scheduled retraining can be useful. For instance, for a sales forecasting model, monthly updating with the latest data helps ensure that the model remains up to date without considerable manual intervention.

Performance-Based Retraining

Another strategy is performance-based retraining. In this case, the model is retrained based on some specific thresholds or metrics. For instance, if the model's accuracy falls below a certain percentage, retraining is triggered to restore the model's performance. Scheduled and performance-based training are the most common strategies; however, we still have more strategies for training to discuss.

Real-Time Retraining

Real-time retraining is used in cases where the data changes very rapidly, and there is a need for the model to adapt continuously. This strategy is resource-intensive but required in dynamic environments, including recommendation systems or stock price prediction, in which user behavior changes rapidly.

Data Drift Detection

Another strategy is retraining the model whenever data drift is detected. By using data drift detection tools such as NannyML or Evidently AI, you can identify when the input data distribution has changed considerably compared to the training dataset. Then, retraining would get triggered.

Strategies for Retraining

Until now, we have investigated why retraining is important and how to find a suitable time for retraining. Here, we discuss the strategies for retraining. Strategies may not be a familiar word for you, so we will clarify that soon. Retraining strategies vary based on the model's data volume, complexity, and computational resources. In this section, you will learn how to select the best strategy for your project. We continue with some well-known strategies for retraining machine learning models.

Full Retraining

In full retraining, the entire model is retrained using both newly collected data and historical data. This approach is used whenever the model's structure needs a fresh start to adapt to big changes. However, it is time-consuming and computationally expensive.

Incremental Learning

Incremental learning is another technique in which the model is updated with just new data without retraining from scratch. This approach is particularly helpful when new data becomes available in small batches and full retraining would be inefficient due to time and cost. Incremental learning is usually used for online learning algorithms, such as support vector machines or decision trees.

Transfer Learning

In deep learning models, transfer learning is used for speeding up the retraining. There, instead of retraining the whole model from scratch, only the final layers are retrained, while the rest of the network remains intact. This technique is helpful when adapting a model to similar but slightly different data.

Example: Using Scikit-Learn for Incremental Learning

In the following, an example of using the `partial_fit` method in `Scikit-Learn` is presented. It simulates initial training and then uses `partial_fit` with a stochastic gradient descent (SGD) classifier for performing incremental learning.

Listing 7-2. An Example of Incremental Learning

```
from sklearn.linear_model import SGDClassifier
import numpy as np
from sklearn.datasets import make_classification
from sklearn.model_selection import train_test_split

# 1. Generate an initial small dataset (Simulate original training data)
X_initial, y_initial = make_classification(n_samples=50, n_features=2,
                                            n_informative=2, n_redundant=0,
                                            random_state=42)
```

```
# 2. Initialize and Train the Model (The first "full" training)
model = SGDClassifier(max_iter=1000, tol=1e-3, random_state=42)
# The classes parameter is essential for the first partial_fit call or
# if you are retraining on an existing model that hasn't seen all classes.
Model.partial_fit(X_initial, y_initial, classes=np.unique(y_initial))

# 3. Simulate New Incoming Data Batches for Incremental Update
X_new = np.array([[0.5, 0.3], [0.9, 0.8], [0.1, 0.2]])
y_new = np.array([1, 0, 1])

# 4. Perform Incremental Learning (Retraining)
# The model is updated with new data without retraining from scratch.
# We don't need to specify 'classes' here as it was done in the
initial fit.
Model.partial_fit(X_new, y_new)

print("Model has been incrementally updated with new data.")
```

In the above example, the model is first initialized with a sample database, then it is updated with new data (`X_new` and `y_new`) without requiring retraining from scratch. This makes the retraining computationally efficient. After talking about retraining strategies, it is time to talk about tools.

Expected Results

```
Model has been incrementally updated with new data.
```

Table 7-1 shows a summary of different retraining strategies.

Table 7-1. *Retraining Strategies Overview*

Strategy	Description	Example Use Case	Advantages	Challenges
Full Retraining	Retrain the model from scratch.	Annual retraining of sales prediction.	Ensures a completely updated model; no residual bias.	Resource-intensive; time-consuming for large datasets.
Incremental Learning	Update the model with new data incrementally.	Online learning in recommendation systems.	Efficient with new data; low resource usage.	May suffer from catastrophic forgetting or bias accumulation.
Transfer Learning	Retrain only part of the model.	Fine-tuning a pre-trained image classifier.	Reduces training time; effective for small datasets.	Requires high-quality pre-trained models; limited to compatible domains.
Active Learning	Selectively retrain on uncertain or mislabeled data.	Retraining on edge cases in fraud detection.	Reduces labeling effort; improves model accuracy.	Needs robust uncertainty quantification.
Federated Learning	Retrain models across decentralized devices.	Collaborative training in healthcare across hospitals.	Preserves data privacy; scalable across networks.	Requires communication efficiency; handles data heterogeneity.
Model Ensembling	Combine retrained and older models for predictions.	Updating ensemble models for stock price predictions.	Improves robustness; captures multiple perspectives.	Computational overhead; complexity in integration.
Fine-Tuning	Adjust specific layers of the model.	Updating language models with domain-specific texts.	Retains existing knowledge; quick to execute.	Requires expertise in selecting layers to update.

(*continued*)

Table 7-1. (*continued*)

Strategy	Description	Example Use Case	Advantages	Challenges
Hybrid Updating	Combines incremental learning with periodic retraining.	Ecommerce inventory demand forecasting.	Balances efficiency and robustness.	Needs careful scheduling and integration.
Bayesian Updating	Updates model parameters using prior distributions.	Probabilistic models for weather forecasting.	Handles uncertainty; works well with probabilistic frameworks.	Limited to specific model types; computationally intensive.
Reinforcement Learning Updates	Adapt policy or value function with new data.	Updating a robot's navigation strategy in real time.	Dynamic adaptability: optimizes performance over time.	Computationally complex; requires high-quality reward signals.

Tools for Automating Retraining

The retraining approach can be automated using different platforms and tools that make the continuous integration and deployment of machine learning models easy. In the following, we discuss some of these tools.

MLflow and Airflow

The first item is MLflow. MLflow can be combined with Apache Airflow to create a complete retraining pipeline for your project. Airflow helps in orchestrating the workflow, ensuring the health of the whole process, including fetching data, retraining the model, evaluation, and then deploying if the retrained model outperforms the current version.

Kubeflow Pipelines

Another tool is Kubeflow. Kubeflow Pipelines provides an easy way to manage and automate retraining workflows on Kubernetes. By using Kubeflow, you can create an end-to-end workflow in which you periodically train and deploy models and provide resilience and scalability.

AWS SageMaker

Amazon SageMaker offers built-in functionality to schedule retraining jobs. You can configure a training job to trigger retraining based on some specific events (such as data drift detection) or even periodically. Based on my experience, explaining the topic with examples helps the users better understand. When I was teaching a difficult topic in university, I was combining that with numerous examples, videos, and some sense of humor to facilitate learning. So, let's consider an example to solidify what we learned here. Assume you have developed a sentiment analysis model for analyzing social media comments. Over time, sentiment expressions and the vocabulary on social media evolve. To keep the model accurate, you should monitor performance using tools such as Prometheus and monitor metrics like precision and accuracy. Also, you can use Evidently AI to detect data drift in the incoming social media comments. After detecting the drift, trigger retraining. When performance falls below a threshold or data drift is detected, it triggers a retraining job. And finally, after successful retraining and validation, you (or your automated pipeline) deploy the updated model. I highly suggest using a CI/CD pipeline. In Figure 7-1, a typical workflow for retraining and updating a deployed machine learning model is shown.

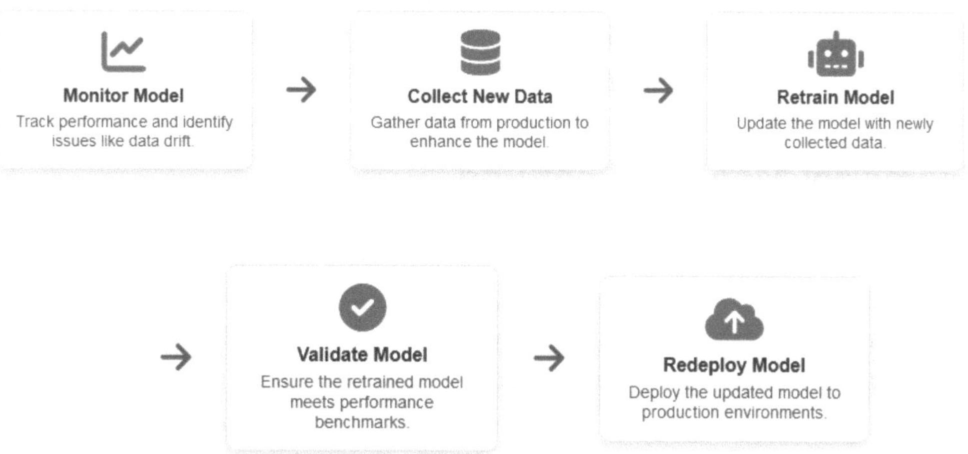

Figure 7-1. *Retraining and Updating Workflow for Machine Learning Models*

Challenges in Retraining and Updating Models

Like all other parts of the ML workflow, there are challenges for retraining and updating the models. The first challenge is data management. You should ensure that the most relevant and correct data is used for retraining. This can be challenging, particularly in large datasets. Another challenge is resource constraints. Retraining needs computational resources, which could be costly and time-consuming, especially for large models. Another challenge is validation and testing. Before deploying an updated model, complete testing is needed to ensure that the model performs better than the previous version. Automated tests and validation pipelines are required to decrease deployment risks.[40]

For effective management of the models, you should make the models secure. In the next section, we will talk about security.

Threat Modeling and Risk Assessment

Security is a critical aspect of any machine learning system. ML systems are prone to a range of threats that can compromise data integrity, functionality, and overall system trustworthiness. Threat modeling and risk assessment involve detecting potential security threats, assessing the risks associated with these threats, and defining appropriate measures to reduce them. This section concentrates on understanding the threats to machine learning systems and assessing vulnerabilities in ML pipelines. If you want to manage your models in production, update them securely, and ensure they are working fine, do not lose this section.

Adversarial Machine Learning

Machine learning models are usually vulnerable to adversarial attacks. These types of attacks exploit the weaknesses of models by giving inputs particularly designed to make the model fail. Adversarial attacks can cause significant risks, specifically when machine learning is used in crucial applications, like financial services, healthcare, and autonomous vehicles. Adversarial machine learning (AML) is a subfield that targets identifying, understanding, and mitigating these vulnerabilities.[31]

Machine learning models, especially deep learning models, have shown considerable capabilities in tasks including natural language processing, image recognition, and reinforcement learning. However, the vulnerability of these models to adversarial manipulation restricts their reliability in practical deployments, particularly in safety-critical systems.[41] Here, we comprehensively talk about adversarial machine learning. We start with the basics of adversarial attacks and continue with different strategies for protecting your system.

The main purpose of adversarial machine learning is to detect, understand, and mitigate the risks caused by adversarial attacks. By testing how adversaries craft deceptive examples, practitioners and researchers can improve the reliability and robustness of machine learning systems. We continue our discussion by investigating different types of adversarial attacks.

Types of Adversarial Attacks

In this section, we discuss various types of adversarial attacks. Also, we will discuss how we can mitigate the effects of each.

Evasion Attacks

Evasion attacks are designed to deceive a model during the inference phase by a small modification of input data. These small changes are imperceptible to humans but are enough to mislead the model into providing wrong predictions. These types of attacks are especially effective against models in classification tasks, including image recognition.[26] For instance, a small change in the pixel values of an image can result in the neural network misclassifying a stop sign as a speed limit sign. This can have serious consequences in autonomous driving systems. Evasion attacks can be successful as some models lack the required generalization to distinguish between genuine inputs and slightly perturbed ones. There are three main methods for crafting invasion attacks:

Fast Gradient Sign Method (FGSM): One of the most common techniques for creating adversarial examples is FGSM. Using the gradient of the loss function with respect to the input, FGSM detects the optimal direction for changing the input to maximize the model's loss.

Projected Gradient Descent (PGD): PGD is an iterative technique for generating adversarial examples. It uses several small perturbations rather than one single large perturbation. This procedure makes PGD even stronger than FGSM at breaking models.

Jacobian-Based Saliency Map Attack (JSMA): JSMA uses the Jacobian matrix to detect features in the input that, when changed, will have the biggest impact on the model's output. This attack can make subtle modifications that are effective in fooling the model.[42] Figure 7-2 illustrates an example of an evasion attack in which an imperceptible change to an image has caused a deep learning model to misclassify it.

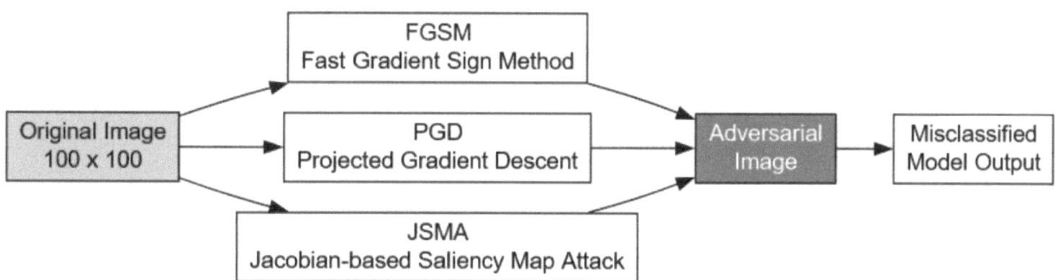

Figure 7-2. *Example of Evasion Attack*

Poisoning Attacks

Poisoning attacks happen during the training phase of a machine learning model. In this type of attack, attackers inject malicious data into the training dataset, which affects the model's decision-making process. Poisoning attacks can reduce a model's accuracy or make it behave unexpectedly on some inputs. So you should be careful that your data is clean and only authorized people have access to it.

A good example of a poisoning attack is changing the training data of a sentiment analysis model by adding a huge number of positive reviews labeled as negative. This can affect the model's understanding of what negative sentiment means. Poisoning attacks are especially important because they can be hard to identify and evade when the model is deployed. Imagine that you have spent a considerable amount of time and money making a model, and finally, you see that because of data poisoning, your model is useless, so painful. So, you should take care of your data properly to prevent this condition. There are several types of poisoning attacks that we will discuss here.

> **Label Flipping**: Label flipping includes changing the labels of training samples. For example, in a binary classification task, the attackers might change the label of some samples from one class to the other class. This causes the model to learn incorrect relationships. Label flipping is making the data incorrect.
>
> **Backdoor Attacks**: It is another type of poison attack in which the attacker poisons the training data with specific patterns. When this specific pattern is present, it causes the model to output a predetermined result. These attacks are especially dangerous as they allow attackers to trigger specific behaviors at inference time. Remember that a backdoor attack can be more complicated and harder to discover from label flipping.

Membership Inference Attacks

The purpose of membership inference attacks is to find out if a specific data sample was part of a model's training set. This causes a significant privacy issue, particularly in sensitive domains like healthcare. This attack is dangerous because identifying whether a person was part of a clinical study can reveal private information.

Technically, membership inference attacks abuse the overfitting of models. In the case that a model overfits to its training data, the model's confidence in classifying training samples may be higher compared to new inputs. This makes it possible to infer if a specific data point was included during training or not.

Model Inversion Attacks

Model inversion attacks try to reconstruct sensitive input data from model outputs. For instance, an attacker can abuse the confidence scores of a facial recognition model to reconstruct the face of a person in the training set. This class of attacks can put the privacy of users at risk and reveal sensitive information about them. Until now, you have gained a good knowledge about various attacks, so for practice, compare membership inference attacks and model inversion attacks.

Implications of Adversarial Attacks

The problems of adversarial attacks are far beyond technical challenges. Attacks on machine learning models can have considerable consequences in various sectors. For example is for the cases of adversarial manipulation in models that are used in financial services. This can lead to incorrect decisions about fraud detection and credit scoring, causing financial losses. Attackers can use adversarial methods to manipulate transaction data and escape fraud detection systems, which can clarify how significant economic damage can result.

In this section, we can talk about some real-world scenarios of adversarial attacks.

Adversarial Attacks on Image Classification

Image classification models are the most susceptible models to adversarial attacks. In these models, imperceptible perturbations can cause models to have wrong classifications. In a security system, such as facial recognition, attackers can use adversarial methods to avoid detection or to impersonate someone else. We had some other examples of image perturbation earlier.

For instance, a small perturbation to an image can result in a model mistakenly identifying a person's face instead of another's, giving unauthorized access to secure areas. Defending against such attacks needs a combination of preprocessing techniques, adversarial training, and continuous monitoring of model outputs.[30] Or in healthcare, adversarial attacks can affect diagnostic models. It is clear how it can be dangerous as it will lead to wrong diagnoses and will put patients' lives in danger. For instance, a small perturbation in medical imaging data can cause a benign tumor to be classified as malignant, which results in unnecessary treatment.

Adversarial Examples in Natural Language Processing (NLP)

In NLP applications, adversarial attacks can alter text inputs in ways that change the intent or meaning without affecting readability. For example, a chatbot could be fooled into returning inappropriate responses by slightly changing the user input. This harms the reputation of the deploying organization. In NLP, adversarial examples include text changes that keep the meaning of a sentence while fooling a model. For example, adding typos or changing words to synonyms can cause NLP models to make wrong classifications or predictions.

These attacks can be especially harmful in machine translation, sentiment analysis, and content moderation applications. For instance, an attacker can slightly modify user reviews to make sentiment analysis models wrongly classify positive reviews as negative, resulting in skewed customer insights.

There are some techniques for defending against adversarial attacks in NLP. Regularizing word embeddings is a good way to decrease the NLP models' sensitivity to small perturbations, including synonym substitutions. Also, training models with character-level variations, like misspellings or typos, can improve robustness to adversarial text modifications. However, you should be careful and training with typos will not result in inaccurate and wrong predictions. Additionally, preprocessing techniques, including spelling correction and synonym replacement, can help in removing adversarial perturbations before feeding the text to the model. My personal preference in real-world projects is preprocessing techniques; however, in some cases, there is a need to use other techniques. Later in this chapter, we will discuss defense mechanisms in more detail.

Adversarial Attacks in Autonomous Systems

Autonomous systems, like self-driving cars and drones, rely on deep learning models for interpreting their environment. Adversarial attacks can have significant consequences in those scenarios. This can lead to lane following errors, incorrect object detection, or other critical misjudgments. For example, a tiny change in the appearance of a road sign can cause an autonomous vehicle to misinterpret its meaning and lead to dangerous driving behavior. These scenarios signify the requirement of strong adversarial defenses in safety-critical systems.

Adversarial Attacks on Reinforcement Learning (RL)

Reinforcement learning models, which learn through interaction with the environment, are also at risk of adversarial attacks. Attackers can manipulate the rewards, the environment, or the observations to make suboptimal behavior.

For instance, in a robotic control system, adversarial perturbations to a sensor can cause the agent to move incorrectly. This can result in damage or accidents. Defending against adversarial attacks in RL is complicated, as the dynamic nature of the environment makes anticipating all possible adversarial scenarios impossible. As an engineer designing the architecture of a machine learning system, recognizing various types of attacks is required. I hope this section has given a good insight into how attackers can put your system in danger and how you can neutralize their attack.

Responsible AI Development

The ethical aspects of adversarial machine learning are considerable. We know that these attacks can exploit vulnerabilities in systems and impact human lives. It is critical to implement frameworks for the responsible development and deployment of machine learning models. These frameworks should address different aspects of responsible AI. There are several aspects of responsible AI such as fairness, transparency, accountability, and also ethical use of data. Privacy and security should be treated as one of the most important topics in the development process. Addressing bias and ensuring a fair outcome is another crucial aspect of responsible AI. Also, there is a need to prioritize sufficient research for adversarial robustness. In the next section, we will talk more about responsible AI development.

Organizations and developers need to have responsible AI development practices, including adversarial testing, regular security assessments, and ongoing monitoring. In addition, ethical guidelines should address the potential misuse of adversarial techniques. By misuse, I mean using them for malicious purposes rather than for enhancing system robustness.

Transparency in model development, such as clear communication about the model's limitations, potential biases, and vulnerabilities, is necessary. Users should be aware of the risks related to using machine learning models, especially in high-stakes domains. Accountability mechanisms must be established to make sure that the organizations deploying machine learning systems are responsible for all harms caused by bias, model failures, or adversarial vulnerabilities.

Defense Mechanisms

There are several defense mechanisms for protecting machine learning models from adversarial attacks.

Adversarial Training

Adversarial training is one of the most effective defenses for adversarial attacks. In adversarial training, a model is trained on both normal data and adversarial examples. This method makes the model more resilient to adversarial perturbations. Because it exposes the model to a variety of adversarial inputs while training, adversarial training needs to generate adversarial samples for every batch during training, which is computationally expensive. In addition, adversarial training is usually specific to the attack type. This means that models trained to resist FGSM attacks are still vulnerable to more sophisticated attacks such as PGD or C&W.

Defensive Distillation

Defensive distillation includes training a model to output probability distributions (i.e., "soft labels") instead of hard classifications. This technique helps make the model less sensitive to small input changes and thus less vulnerable to adversarial attacks. In defensive distillation, first, the model is trained using the original data. In this stage, the output probabilities are used as soft labels for training a distilled model. This procedure makes the decision boundaries smoother, which results in reducing the model's susceptibility to adversarial perturbations.

Input Preprocessing

Input preprocessing methods include transforming the input data in a way that filters out adversarial perturbations. Methods such as feature squeezing, image compression, and input denoising are good practices to remove adversarial noise before feeding the data to the model. Feature squeezing decreases the precision of input features and so reduces the number of degrees of freedom that are available to an adversary. For example, decreasing the color depth of an image can eliminate the small perturbations that adversaries have added. In addition, using denoising autoencoders that can preprocess inputs by learning to remove noise can be beneficial. These noises include adversarial perturbations. If you have experience in SQL, you can compare input preprocessing with sanitizing SQL queries to prevent SQL injection.

Randomized Smoothing

Randomized smoothing is a method for converting a classifier to a smoothed classifier. This method adds random noise to the input before making predictions. Randomized smoothing provides provable robustness against adversarial attacks.

Gradient Masking

Gradient masking is a defense strategy. The purpose of gradient masking is to reduce the usefulness of the model's gradient information to an adversary. This method hides or obfuscates gradient information, which hinders the generation of adversarial samples using gradient-based methods. However, this approach has restrictions, as it can be circumvented by more advanced attacks that do not rely only on gradient information. Table 7-2 summarizes the defense mechanism against adversarial attacks.

Table 7-2. *Defense Mechanisms Against Adversarial Attacks*

Defense Mechanism	Description	Example Use Case
Adversarial Training	Train on adversarial examples to build robustness	Image classification in autonomous vehicles
Defensive Distillation	Use soft labels to reduce sensitivity	Enhancing model security in healthcare
Input Preprocessing	Apply data transformations to remove perturbations	Preventing attacks in computer vision
Randomized Smoothing	Add noise to inputs to provide robustness	Improving model stability in NLP tasks

Assessing Vulnerabilities in ML Pipelines

A machine learning pipeline consists of several parts, such as data collection, data preprocessing, model training, and model deployment. All these components have some unique vulnerabilities that must be assessed to secure the entire system efficiently. In this part, we talk about the main vulnerabilities in ML pipelines.

Data Collection and Preprocessing: This component contains data integrity and validation. The data used in training should be collected from trustworthy sources to ensure data integrity. Manipulated datasets or data from untrusted sources lead to poisoning attacks or can introduce bias. Also, it is important to validate the data before using it in training. Techniques such as data normalization and outlier detection can help in identifying and mitigating the risks associated with malicious or untrusted data points.

Model Training: In model training, we should consider access control while training on confidential data. During the model training phase, unauthorized access to computing resources and training data can compromise the model. So, user authentication and implementing strict access control policies are important for protecting the training environment. Additionally, if the training includes personally identifiable information (PII) or confidential data, encryption or privacy-preserving techniques, like differential privacy, should be used to save sensitive data from leakage.

Model Deployment: In this part, API security and model versioning are important. After a model is deployed, it is usually exposed via an API for inference. API security is critical for preventing unauthorized access. Techniques such as authentication, rate limiting, and encryption help in securing the API endpoints. Keeping track of various model versions is necessary for risk assessment. As an attacker may try to exploit vulnerabilities in an outdated model version, it is important to manage and update deployed models systematically.

Risk Assessment Process

The following steps help assess vulnerabilities and reduce risks in ML systems. I highly recommend considering them in your projects.

1. **Identifying Assets**: Identify the crucial assets in your machine learning pipeline, such as APIs, data, models, and infrastructure.
2. **Identifying Threats**: Identify potential threats that can affect the identified assets. Such as data poisoning, model inversion, and all other attacks discussed earlier.
3. **Determining Vulnerabilities**: Determine the vulnerabilities in the ML pipeline, such as the lack of encryption, unsecured data sources, and insufficient access control mechanisms.
4. **Analyzing Impact**: Analyze the impact of the known threats. How would a successful attack impact the data integrity, functionality, or privacy of your machine learning system?
5. **Mitigating Risks**: Define security measures to reduce the identified risks. Such as encryption, implementing data validation, access controls, and regular security audits.

Figure 7-3 shows threat modeling in an ML pipeline. It highlights key assets, potential threats, and mitigation strategies. It detects threats at different stages, including data collection, training, and deployment, and represents how to secure machine learning workflows.

***Figure 7-3.** Threat Modeling in Machine Learning Pipelines*

Figure 7-3 shows the main components of an ML pipeline and detects the potential threats at each stage, providing a clear understanding of how to secure machine learning workflows. Additionally, for effective defense, we can implement outlier detection algorithms. These algorithms can help identify and remove the data points that are suspicious of a poisoning attack. Techniques like Local Outlier Factor (LOF) and Isolation Forest are usually used to detect anomalies in the dataset. Practice maintaining data provenance; for that, keep track of the origin of all data points for data authenticity. This process facilitates tracking of compromised training data.

In addition to the above techniques, model hardening techniques make the model more resistant to adversarial attacks. They modify the training process or model architecture. Here, we talk about two model hardening techniques: gradient masking and ensemble models. In gradient masking, the model's gradients are obscured, which makes it difficult for attackers to use gradient-based techniques such as FGSM to generate adversarial examples. This technique should be used cautiously, because it may lead to obfuscation rather than true robustness. Ensemble models use several models in an ensemble to decrease the probability of successful attacks. By using ensemble methods, adversarial examples that are made to target a specific model might not transfer to all models in an ensemble.

Again, limiting access to the model and the training dataset is a basic way to reduce risks. It is not offensive for employees; it is required for business. By implementing strict access control policies, only authorized personnel can update the model or training data or make changes to the model. Remember that applying digital signatures to the data and model helps ensure that any changes made can be detected. If an adversary tries to modify a model, the absence of a valid signature alerts the system to potential tampering. Figure 7-4 shows the types of adversarial attacks in machine learning, such as evasion and poisoning, and their corresponding defense mechanisms. This figure highlights techniques such as data validation, adversarial training, and model hardening to enhance system resilience and reduce risks.

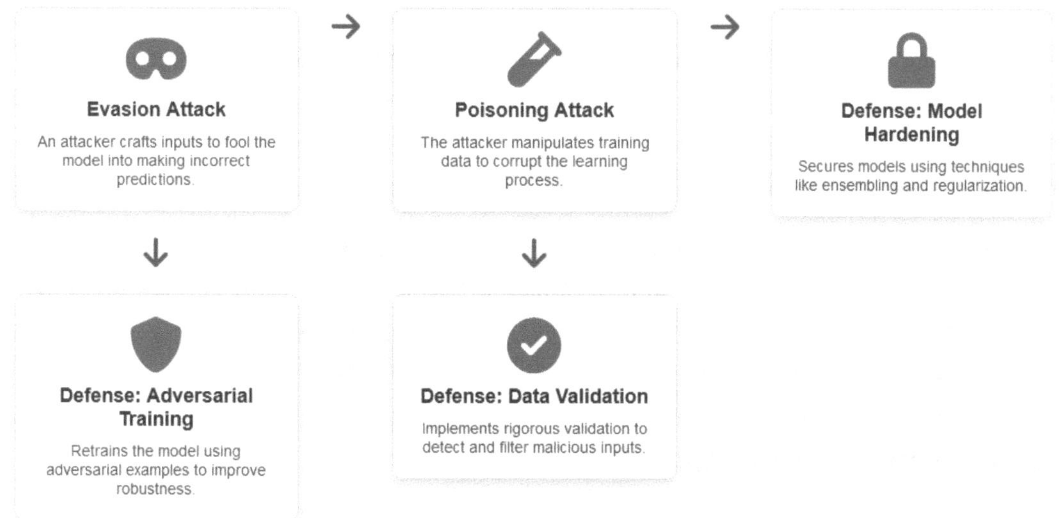

Figure 7-4. *Adversarial Attacks and Defense Techniques in Machine Learning*

We talked about various attacks and their defense strategy. Table 7-3 shows a summary of adversarial attacks and the defense methods.

Table 7-3. *Overview of Adversarial Attacks and Defenses*

Attack Type	Description	Defense Strategy
Evasion Attack	Crafting perturbed inputs to fool the model	Adversarial training, gradient masking
Poisoning Attack	Injecting malicious data during training	Data sanitization, outlier detection
Model Inversion	Inferring sensitive information from outputs	Differential privacy, limiting API access
Model Stealing	Reconstructing the model by querying it	API rate limiting, response noise

Data Security and Protection

Data protection and security are important components of machine learning systems. By knowing the sensitive nature of training data, it is necessary to implement robust practices for confidentiality, data integrity, and protection from unauthorized access.

Here, we focus on the best practices for securing training data and the techniques for secure storage and data encryption that help in reducing the risks of data breaches and other security threats.

Best Practices for Securing Training Data

Training data is the heart of machine learning models. Ensuring data security is required for building trustworthy and reliable models. By implementing the following best practices, you can significantly decrease the risk of unauthorized access and data exposure.

Access Control and Authentication

As we mentioned earlier, the most important practice is limiting access for the user. Access to training data must be restricted based on responsibilities and roles. Implementing strict access control policies guarantees that only authorized personnel can modify or have access to the training data. For this practice, consider implementing Role-Based Access Control (RBAC). RBAC is a well-known method for managing user permissions. Engineers, data scientists, and administrators should have various access levels based on their roles. For example, data scientists may have read-only access while administrators may have permission to modify data. Also, Multi-Factor Authentication (MFA) is another good practice. By enforcing MFA, you add an extra layer of security to the access control process. This layer hardens, making unauthorized access for attackers. All employees should know this structure and will not get offended because of their limited access. We should remember that business is not a family; business is business; it is a contract between employees and employer, and its purpose is to make a product. It is the right of the owner of the model or data to keep their product secure, and employees should understand it.

Data Masking and Anonymization

Anonymization and data masking are crucial techniques for protecting sensitive data such as personally identifiable information (PII) during model training. Data masking includes obfuscating sensitive information by replacing it with fictitious but realistic data. Masking makes sure that during training, while preserving the data's utility for analysis, real data will not be exposed. Another technique is **data anonymization**. Data anonymization includes altering or removing data identifiers, like addresses and names, to prevent the re-identification of individuals. This approach helps compliance with privacy regulations, like the GDPR, and at the same time maintains data utility for training purposes. Assume we have medical data of some patients, and we want to use them for training a model for disease diagnosis. We can remove personal data (name, address, etc.) from the data before using it in model training. It is obvious that personal data does not have any effect on our diagnosis model. However, anonymizing the data keeps the privacy of patients.

Data Integrity Verification

Another practice for data security is data integrity verification. Data integrity is crucial for ensuring that training data has not been manipulated. By implementing mechanisms for data integrity verification, you can enhance reliability and detect unauthorized changes in the training process. One method for ensuring data integrity is using checksums and hash functions. Hash functions or checksums make any modifications to the data detectable. In this case, if the calculated hash value of the data changes, it shows that the data has been modified. Another technique is using digital signatures. Digital signatures can be applied to data files to ensure their integrity and authenticity. In the case that a data file is altered, the digital signature would not be valid anymore, which provides a clear indication of tampering. You may have heard about blockchain chain which is a very secure technique in cryptocurrencies. If you are interested, you can do some research about it and see how hashing has resulted in such a secure technique.

Secure Data Sharing

In practical environments, training data is usually shared among organizations or teams. For these cases, implementing secure data-sharing mechanisms is needed to maintain data integrity and confidentiality during transit. It is recommended to encrypt data in transit. Encrypting data before sharing ensures the security of sensitive information during transmission. Techniques such as Transport Layer Security (TLS) can be used to secure data in transit. Another good practice is considering Access Expiry. Data sharing should be time-bound and with access permissions that should expire after a specified period. This practice prevents unnecessary or prolonged exposure of sensitive data. To give you some clues, consider that HTTPS is secure, authentication is based on a token, and tokens have a limited life. I wish I could explain everything here, but if I do so, the book would be 10,000 pages!

Techniques for Data Encryption and Secure Storage

Encrypting data and secure storage are required to ensure the integrity and confidentiality of training data. Data encryption both in transit and at rest helps protect against unauthorized access and data breaches. When possible, we can encrypt all data; in some cases of super huge data, frequent encrypting and decrypting are not practical. When I was teaching at university, students were asking which types of data should be encrypted and which should not. My short answer was that you assume that the whole world sees the data that you do not encrypt. Now, if it is a matter, you should encrypt them. Here we talk more about encryption.

Encryption for Data at Rest

Data stored on physical media is called data at rest. Data at rest can rest in databases, hard drives, or cloud storage! Encrypting data at rest ensures that, in the case that an attacker has access to the storage medium, they could not read the data without the encryption key. Attacker access to physical devices is not impossible. In IoT devices, attackers can steal devices; additionally, stealing a hard drive from an office or even a data center is not impossible. There are several encryption methods; one method is AES (Advanced Encryption Standard). AES is a common encryption standard for securing data at rest. It offers strong encryption with different key sizes, including 128, 192, or 256 bits. Also, do not forget to encrypt your database. Built-in database encryption features like Transparent Data Encryption (TDE) can help in encrypting databases storing training data. TDE automatically encrypts all data within the database without needing application-level changes. Be aware that although encryption offers considerable security, it generates an extra load on your machines, too.

Encryption for Data in Transit

Data that is being transmitted between systems or components is called data in transit. This data includes the data that is transferred between a client and a server. Encryption for data in transit reduces the risk of unauthorized entities reading or intercepting sensitive information during transmission. TLS is the de facto standard for encrypting data in transit. It offers a secure channel for communication by encrypting the data exchanged between the client and server. In addition, using Virtual Private Networks (VPNs) is also useful. VPNs can create secure communication channels, specifically while transmitting data between cloud environments and on-premise infrastructure.

Secure Key Management

Encryption is strong, just in case of good protection of the encryption keys. Secure key management practices are necessary for keeping the security of encrypted data. One effective way for protecting keys is Hardware Security Modules (HSMs). HSMs are hardware devices that are used for securely generating, storing, and managing encryption keys. They give a high level of protection against unauthorized access. Another method is Cloud Key Management Services. Cloud providers such as Azure, AWS, and Google Cloud provide managed key management services. These services offer secure key generation, storage, and rotation. Using managed services decreases the operational burden of secure key management.

Backup Encryption

I have seen many cases where organizations keep data in rest and transit secure, but do not care about their backups. They think, by default, backups are safe! Regular backups are necessary for data security. On the other hand, backup data can be a target for attackers, too. Encrypting backup files is required to keep the data secure even if the backup media is compromised. Using backup tools that support encryption is a good practice for automatically encrypting backup files. For instance, AWS Backup can be used to back up services of AWS resources. AWS Backup has built-in support for encryption. Figure 7-5 shows different techniques for data security in machine learning systems, such as encryption, data masking, and secure data storage. It illustrates the importance of securing training data through encryption, integrity verification, and access control, which results in a comprehensive security approach.

Figure 7-5. *Data Security Techniques in Machine Learning*

Secure Model Deployment

Secure deployment of machine learning models is very important in building robust machine learning systems. Models in production are exposed to many potential risks and vulnerabilities. These risks include unauthorized access, data leaks, and attacks targeted at compromising the integrity of the system. To surpass these risks, it is required to implement secure deployment strategies, like authentication, API security, model isolation, and sandboxing.[29]

Here, we discuss the main techniques for ensuring secure model deployment.

API Security and Authentication for ML Models

APIs are the main way for machine learning models to interact with other applications. When you send a request to an LLM on the web and you see the results, you are using the API. This application of API makes them a critical component of model deployment. Also, it is very probable that you need an API for your projects. Proper authentication and ensuring API security prevent unauthorized access and help in maintaining the model integrity.

Authentication and Authorization

OAuth 2.0 is a common protocol for secure access control. It issues secure tokens and guarantees that only authorized services or users can access the model API. Tokens are assigned based on user credentials, and their use blocks requests from unauthorized entities. Another tool is API keys. API keys are a simple way to limit access to the model. API keys are offered to authenticated users, and there is a need to be included in all requests for the model. I should mention that it is important to store API keys securely, because compromised keys can result in unauthorized access.

Another technique is using JWT (JSON Web Tokens), which is very popular these days. JWT tokens are usually used for stateless authentication, which allows users to authenticate themselves securely without keeping session data on the server. This method is efficient, particularly for large-scale deployments in which scalability is important. Before we talked about microservices, we mentioned that a microservice is supposed to break a big project into several standalone projects. Using JWT, as it is stateless, is a good choice in a microservices architecture.

Rate Limiting and Quotas

You should consider rate limiting on the model API to manage resource usage efficiently and prevent Denial-of-Service (DoS) attacks. Rate limiting limits the number of client requests in a specific time range and blocks automated API calls that are too frequent. Be careful about setting the rate limit so as not to unnecessarily block your users. API gateways like NGINX or AWS API Gateway can set rate limits. Using these tools helps in managing the client requests and prevents overloading the model server.

HTTPS for Secure Communication

It is necessary to encrypt communication between the model API and clients to prevent data interception. If you do not encrypt your data while transferring, data can be accessed or tampered with. HTTPS (Hypertext Transfer Protocol Secure) should be used to protect the integrity and confidentiality of data in transit. TLS certificates can be used for establishing secure connections between clients and the model API. Cloud services, like AWS Certificate Manager, provide an easy way to manage and deploy TLS certificates for your API.

Logging and Monitoring

Logging all API requests, such as access failures, authentication attempts, and resource usage, helps in finding suspicious behavior and security incidents. Tools such as Splunk or ELK Stack (Elasticsearch, Logstash, Kibana) can be used to collect and analyze logs from the API server. These tools enable quick detection and mitigation of potential threats. So, here we have talked enough about API security. In the next section, we talk about the security of the deployed model.

Model Isolation and Sandboxing Techniques

Model isolation and sandboxing are methods used to protect the deployment model from potential attacks or unauthorized manipulation. In this section, we talk more about model isolation techniques.

1. Containerization for Model Isolation

We had an extensive talk about containerization before. There, we discussed how containerization can help in modularity and scalability. Here, we discuss the containers from a security point of view. Containerization is a well-known approach for isolating models from other processes and services. By deploying each model in its container, you can operate the model in a controlled environment and prevent it from affecting other system components if compromised.

In scenarios where stricter isolation is needed, you can deploy models in separate virtual machines (VMs). VMs offer stronger isolation in comparison to containers. Since each VM has its own operating system, it would be more challenging for the attackers to gain access to other services if they succeed in compromising the VM. Cloud providers such as Azure Virtual Machines, AWS EC2, and Google Compute Engine help users to deploy models on isolated virtual machines, which enhances their security for sensitive use cases. Again, I emphasize that in highly secure cases, virtual machines are used rather than containers.

2. Sandboxing Techniques

Sandboxing means running code in an isolated environment with limited access to system resources. This approach helps reduce the risks of running untrusted or potentially malicious code. You can deploy your models in a sandboxed environment where access to file systems, system resources, and network interfaces is restricted. This approach prevents models from affecting the host system in case they are compromised.

Tools like **gVisor** and **Firejail** are popular for sandboxing in Linux environments. **Firejail** uses Linux namespaces to prepare a sandbox for processes; on the other hand, **gVisor** is an open source project that has an additional layer of isolation between the host kernel and the container. In some companies, sandboxing is mostly used in the staging environment, too. A staging environment is a duplicate of the production environment used for testing code before publishing it to the public. In several companies, I have seen this situation.

3. Resource Limits

Setting resource limits on memory, CPU, and disk usage for the model is a good practice because even if the model gets compromised, it cannot disrupt other processes by exhausting system resources. Kubernetes gives administrators the ability to define resource quotas for each pod. By this, no single model deployment can consume excessive resources. I remember one company asked me to check their system robustness, and after having a quick check, the first question was, "Have you limited resource usage for different services?" I asked this question because limiting resources is very important. If you do not limit the resources of the services in a shared environment, a compromised service can easily disrupt all services that are using shared resources. Figure 7-6 represents secure model deployment strategies, such as containerization, API security, and sandboxing. It emphasizes components like model isolation, secure communication, and resource control for effective mitigation of deployment risks. These strategies are the backbone of a robust system.

Figure 7-6. *Secure Model Deployment Strategies*

A summary of secure development techniques is listed in Table 7-4.

Table 7-4. *Overview of Secure Deployment Techniques*

Technique	Description	Tools/Technologies
API Security and Authentication	Protecting model APIs from unauthorized access	OAuth 2.0, API keys, JWT, HTTPS
Containerization	Isolating models using containers	Docker, Kubernetes
Sandboxing	Running models in restricted environments	Firejail, gVisor
Resource Limits	Limiting resource usage to prevent overconsumption	Kubernetes Resource Quotas

Project Title: Managing and Updating ML Models in Production

In this chapter, we discussed managing and updating ML models. Here, we can have a simulation to practice our learning.

Objective

This project shows a complete simulation for versioning, retraining, managing, and monitoring machine learning models in production. Here, we use a real publicly available dataset for breast cancer. This dataset is used in many researches and for production model training. In this project, you will learn how to register model versions, apply incremental retraining, and monitor model performance over time. This simulation emphasizes key concepts such as best practices of model versioning, retraining, and monitoring/alerting in a production environment. We expect that at the end of this project, you can use your knowledge effectively to update a model in production and track its performance and metadata effectively.

Project Description

In this project, you will build a simulation that mimics managing an ML model in production. The simulation has the following modules:

1. **Model Registry and Versioning:** A system for registering and tracking various versions of the model and its metadata.
2. **Incremental Retraining:** An example of updating a model using incremental learning methods.
3. **Monitoring and Alerting:** A monitoring system that tracks performance metrics and triggers alerts whenever performance degrades.

All modules are implemented as a Python class. In this project, outputs are printed at each step.

Step 1: Import Required Libraries

The code below imports the required libraries. Similar to all other projects of this book, in the first step, we load libraries required for our simulation:

- `NumPy`: For data generation and numerical operations.
- `matplotlib`: For visualizing monitoring metrics.
- `SGDClassifier from scikit-learn`: This is for simulating model retraining and incremental learning.
- `accuracy_score`: To measure the model's performance.
- `make_classification`: To load publicly available datasets for training and retraining.
- `train_test_split`: To separate data into train and test.
- `StandardScaler:` For standard scaling.
- `time`: To simulate delays during retraining and registration.

Listing 7-3. Step 1: Import Required Libraries

```
import numpy as np
import matplotlib.pyplot as plt
from sklearn.linear_model import SGDClassifier
from sklearn.metrics import accuracy_score
from sklearn.datasets import load_breast_cancer
from sklearn.model_selection import train_test_split
from sklearn.preprocessing import StandardScaler import time
```

Expected Results

The code of this step should run without producing any output. Remember that in all cases of loading libraries, no result is good news of no error.

Step 2: Define the Model Registry and Versioning Classes

The code below makes a class for model versioning. In this step, we write two classes:

- `ModelVersion`: Keeps information about the model version, such as its version number, timestamp, performance accuracy, and metadata.
- `ModelRegistry`: Simulates a registry for saving and tracking various model versions. It provides methods to register new versions and list all registered models.

Listing 7-4. Step 2: Define the Model Registry and Versioning Classes

```
class ModelVersion:
    def __init__(self, version, model, accuracy, metadata=None):
        """
        Initialize a model version.

        Args:
            version (str): Version number (e.g., "1.0.0").
            model: The trained model object.
            Accuracy (float): Model accuracy on a validation set.
            Metadata (dict, optional): Additional metadata (e.g.,
            training time).
        """
        self.version = version
        self.model = model
        self.accuracy = accuracy
        self.metadata = metadata or {}
        self.timestamp = time.strftime("%Y-%m-%d %H:%M:%S")

    def __str__(self):
        return f"Version: {self.version}, Accuracy: {self.
        accuracy*100:.2f}%, Timestamp: {self.timestamp}, Metadata: {self.
        metadata}"
```

```
class ModelRegistry:
    def __init__(self):
        self.registry = []
        print("Model Registry initialized.\n")

    def register(self, model_version):
        """
        Register a new model version.

        Args:
            model_version (ModelVersion): The model version to register.
        """
        self.registry.append(model_version)
        print(f"Registered model version: {model_version}\n")

    def list_versions(self):
        """
        List all registered model versions.
        """
        print("Listing all model versions:")
        for version in self.registry:
            print(version)
```

Expected Results

When you make an instance of `ModelRegistry` and register a model version, you will see a printed message of the model version details.

Step 3: Simulate Incremental Model Training (Retraining)

The code below simulated the incremental model training. The `IncrementalModel` class uses scikit-learn's `SGDClassifier` to simulate a model that can be updated incrementally. The `initial_train` method does the first training, and the `incremental_update` method simulates retraining the model with the new data. The accuracy of the model is printed after each training step, if done to simulate performance monitoring.

Listing 7-5. Step 3: Simulate Incremental Model Training (Retraining)

```
class IncrementalModel:
    def __init__(self):
        # Initialize the SGDClassifier with partial_fit support
        self.model = SGDClassifier(max_iter=1000, tol=1e-3)
        self.classes_ = np.array([0, 1])  # Binary classification
        print("Initialized Incremental Model.\n")

    def initial_train(self, X, y):
        """
        Train the model initially using a training dataset.

        Args:
            X (numpy.ndarray): Training features.
            Y (numpy.ndarray): Training labels.
        """
        print("Performing initial training of the model...")
        self.model.partial_fit(X, y, classes=self.classes_)
        acc = accuracy_score(y, self.model.predict(X))
        print(f"Initial training complete. Accuracy: {acc*100:.2f}%\n")
        return acc

    def incremental_update(self, X_new, y_new):
        """
        Update the model with new training data using incremental learning.

        Args:
            X_new (numpy.ndarray): New training features.
            Y_new (numpy.ndarray): New training labels.
        """
        print("Performing incremental update on the model...")
        self.model.partial_fit(X_new, y_new)
        acc = accuracy_score(y_new, self.model.predict(X_new))
        print(f"Incremental update complete. New data accuracy:
        {acc*100:.2f}%\n")
        return acc
```

Expected Results

After running the initial training and incremental updates, you will see printed outputs illustrating the training progress. In addition, a printed message shows the model's accuracy.

Step 4: Simulate Monitoring and Alerting

The code below makes a model for simulating the monitoring system. The `MonitoringSystem` class records the model's accuracy in different stages. If the accuracy drops below a predefined threshold (e.g., 80%), it triggers an alert. The `display_metrics` method makes a plot of the recorded accuracy metrics, visually providing a monitor of model performance.

Listing 7-6. Step 4: Simulate Monitoring and Alerting

```
class MonitoringSystem:
    def __init__(self, alert_threshold=0.80):
        """
        Initialize the monitoring system.

        Args:
            alert_threshold (float): Minimum acceptable accuracy.
        """
        self.alert_threshold = alert_threshold
        self.metric_log = []
        print("Monitoring system initialized with alert threshold at
        {:.2f}%.\n".format(alert_threshold*100))

    def record_metric(self, accuracy):
        """
        Record the current model accuracy and check if an alert should be
        triggered.

        Args:
            accuracy (float): The current accuracy of the model.
        """
        self.metric_log.append(accuracy)
        print(f"Recorded model accuracy: {accuracy*100:.2f}%")
```

```
        if accuracy < self.alert_threshold:
            print("ALERT: Model accuracy has dropped below the
            threshold!\n")

    def display_metrics(self):
        """
        Plot the recorded accuracy metrics over time.
        """
        iterations = range(1, len(self.metric_log) + 1)
        plt.figure(figsize=(8, 4))
        plt.plot(iterations, [acc*100 for acc in self.metric_log],
        marker='o', linestyle='-', color='purple')
        plt.xlabel("Iteration")
        plt.ylabel("Accuracy (%)")
        plt.title("Model Accuracy Over Time")
        plt.tight_layout()
        plt.show()
```

Expected Results

After recording several metrics, you can see the printed messages of each recorded accuracy. If accuracy drops below 80%, an alert message is generated and printed.

Step 5: Orchestrate the Entire Model Management and Updating Process

The code below orchestrates the entire model's management system. The `main_production_management()` function combines all components. In this step, the following parts are completed:

- **Initial Training and Registration:** Synthetic data is generated and used to train the first model. This version is registered as version 1.0.0.
- **Incremental Retraining:** New data simulating data drift is generated and updates the model incrementally. The updated model is registered as version 1.1.0.

- **Monitoring:** The monitoring system records accuracy before and after retraining. If accuracy drops below the threshold, an alert is triggered. Finally, a plot illustrates the trend in model accuracy over time.
- **Model Registry Listing:** All registered model versions are listed for verifying the versioning process.

Listing 7-7. Step 5: Orchestrate the Entire Model Management and Updating Process

```
def main_production_management():
    print("=== Starting Model Management and Updating Simulation ===\n")

    # 1. Initialize the model registry, incremental model, and
    monitoring system
    registry = ModelRegistry()
    inc_model = IncrementalModel()
    monitor = MonitoringSystem(alert_threshold=0.90) # Increased threshold
    for a real dataset

    # 2. Load Real Data and Prepare Batches
    print("Loading Breast Cancer dataset for simulation...")
    data = load_breast_cancer()
    X_full, y_full = data.data, data.target

    # Standardize data (crucial for SGDClassifier performance on real data)
    scaler = StandardScaler()
    X_full = scaler.fit_transform(X_full)

    # Split data into Initial Training (60%) and New Retraining Batch (40%)
    X_initial, X_rest, y_initial, y_rest = train_test_split(X_full, y_full,
    test_size=0.4, random_state=42)
    # Take 50% of the rest as the "new" batch for retraining (40% * 0.5 =
    20% of total)
    X_new, _, y_new, _ = train_test_split(X_rest, y_rest, test_size=0.5,
    random_state=24)

    print(f"Initial Training Samples: {len(X_initial)}")
```

```
    print(f"Incremental Update Samples: {len(X_new)}\n")

    # 3. Perform initial training and register version 1.0.0
    initial_acc = inc_model.initial_train(X_initial, y_initial)
    version1 = ModelVersion(version="1.0.0",
                            model=inc_model.model,
                            accuracy=initial_acc,
                            metadata={"dataset": "Breast_Cancer",
                            "samples": len(X_initial)})
    registry.register(version1)
    monitor.record_metric(initial_acc)

    # Simulate waiting period before new data arrives
    print("Waiting for new data for retraining…\n")
    time.sleep(1)

    # 4. Perform incremental update (retraining) and register version 1.1.0
    # Note: SGDClassifier performs poorly if the data changes drastically,
    # so we expect a small performance change, which is realistic.
    New_acc = inc_model.incremental_update(X_new, y_new)
    version2 = ModelVersion(version="1.1.0",
                            model=inc_model.model,
                            accuracy=new_acc,
                            metadata={"dataset": "Breast_Cancer", "new_
                            samples": len(X_new)})
    registry.register(version2)
    monitor.record_metric(new_acc)

    # 5. List all registered model versions
    registry.list_versions()

    # 6. Display monitoring metrics over iterations
    monitor.display_metrics()

    print("=== Simulation Complete ===")

# Execute the simulation if this cell is run directly
if __name__ == "__main__":
    main_production_management()
```

Expected Results

```
=== Starting Model Management and Updating Simulation ===

Model Registry initialized.

Initialized Incremental Model.

Monitoring system initialized with alert threshold at 90.00%.

Loading Breast Cancer dataset for simulation...
Initial Training Samples: 341
Incremental Update Samples: 114

Performing initial training of the model...
Initial training complete. Accuracy: 96.48%

Registered model version: Version: 1.0.0, Accuracy: 96.48%, Timestamp:
2025-12-10 06:49:54, Metadata: {'dataset': 'Breast_Cancer', 'samples': 341}

Recorded model accuracy: 96.48%
Waiting for new data for retraining...

Performing incremental update on the model...
Incremental update complete. New data accuracy: 98.25%

Registered model version: Version: 1.1.0, Accuracy: 98.25%, Timestamp:
2025-12-10 06:49:55, Metadata: {'dataset': 'Breast_Cancer', 'new_
samples': 114}

Recorded model accuracy: 98.25%
Listing all model versions:
Version: 1.0.0, Accuracy: 96.48%, Timestamp: 2025-12-10 06:49:54, Metadata:
{'dataset': 'Breast_Cancer', 'samples': 341}
Version: 1.1.0, Accuracy: 98.25%, Timestamp: 2025-12-10 06:49:55, Metadata:
{'dataset': 'Breast_Cancer', 'new_samples': 114}
```

After running `main_production_management()`, you will see the printed outputs showing the progress of training, retraining, and registration of model versions with their metadata, it also alerts if the model accuracy falls below the threshold. Additionally, a figure (Figure 7-7) that shows model accuracy over time will be generated.

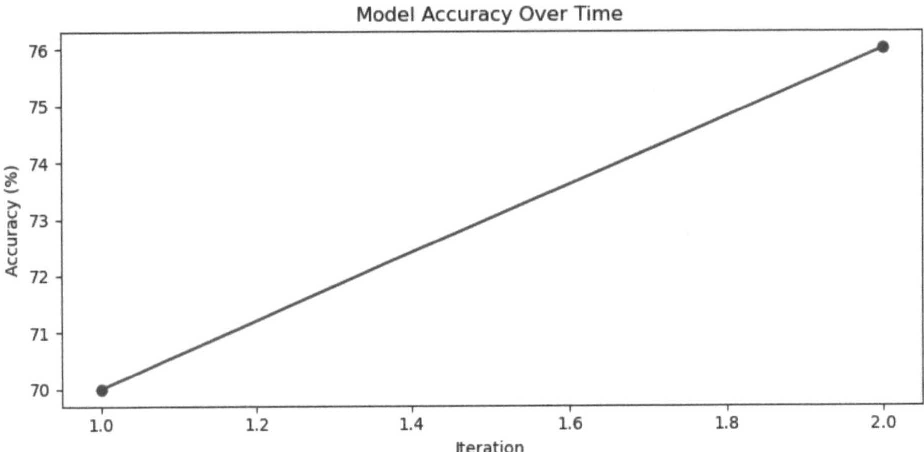

***Figure 7-7.** Model Accuracy Over Time*

Conclusion

This project simulates the updating and management of machine learning models in production. In this simulation, you learned how to register and version models (using a simple model registry). Perform incremental retraining (with synthetic data to handle data drift). Monitor model performance and trigger alerts when performance drops below a threshold, and finally, visually track the model accuracy over time. This project covers the complete cycle of retraining and monitoring.

The modular design of this simulation lets you test your knowledge of best practices in model versioning, retraining strategies, and monitoring in a production environment. Each section is completely designed to run smoothly in a Jupyter Notebook environment, making a standalone, comprehensive guide for managing ML models in production. This project is a good practice to take care of your ML projects in production.

Summary

After deploying the models on production, another phase of machine learning starts. In this phase, you need to maintain your model. You have to monitor it continuously, and if its accuracy degrades, have a solution for that.

In this chapter, we covered the critical aspects of maintaining machine learning models in production. We started by emphasizing the importance of model versioning for ensuring accountability, reproducibility, and seamless updates. Then we proceed with the main best practices, including storing metadata, adopting semantic versioning, and leveraging model registries like DVC, MLflow, and Kubeflow. Afterward, we clarified all principles with some real-world examples.

We also addressed updating and retraining models to ensure continuous performance in the production environment. We talked about scenarios for retraining, such as performance-based triggers, scheduled intervals, real-time updates, and data drift detection. We discussed strategies like incremental learning, full retraining, and transfer learning, and tools like Airflow, MLflow, Kubeflow Pipelines, and AWS SageMaker. We said in which cases we should have full retraining and in which cases incremental retraining is a better choice.

Another important aspect of managing and maintaining machine learning projects is to keep them safe. In this chapter, we discuss various types of attacks, their impact, and defensive scenarios to mitigate the risks. It is required for all developers and managers to know the adversaries and have a plan for them; if not, it can result in bad consequences. In addition to attacks, a wrong architecture and design can have consequences, too. In the next chapter, we discuss how to test and debug our models to ensure its reliability.

CHAPTER 8

Testing and Debugging ML Systems

Ensuring the robustness and reliability of machine learning (ML) models is critical for production-grade applications. Unlike traditional software systems, ML models are stochastic, data-dependent, and evolve. This makes testing and debugging more complicated. Errors in ML systems can arise from different sources, such as model training bugs, data pipeline issues, hyperparameter misconfigurations, and model drift. By implementing a systematic method for testing and debugging ML models, you ensure reproducibility, consistency, and fairness in your project. In addition, you decrease failures in practical deployments. This chapter focuses on best practices for testing and debugging ML systems, such as

Unit Testing for ML Models: This guarantees the correctness of each component, like feature engineering, data transformations, and model predictions.

Debugging Data Pipeline Issues: This includes detecting and resolving problems related to preprocessing, data ingestion, and feature extraction. We had a comprehensive talk about the data pipeline in Chapter 2, and here, we will enhance our knowledge by learning how to test it.

Addressing Model Performance Failures: This includes diagnosing issues related to overfitting, model bias, underfitting, and distributional shifts in real-world data. Again, we have discussed model performance and monitoring before, and here, we will talk about testing it.

M. R. Mahdiani, *Mastering Machine Learning Architecture and Solutions*,
https://doi.org/10.1007/979-8-8688-2527-9_8

Do not forget to implement a robust automated test system in your ML project. It may take some time and energy, but in the long term, it pays back. By implementing comprehensive testing and debugging techniques, you and your team, including data scientists and ML engineers, can reduce errors, improve model reliability, and streamline ML workflows for production deployment. We start the chapter with unit testing.

Unit Testing for ML Models

Unit testing is a basic practice in software engineering. In all software and coding projects, unit testing plays a crucial role. Unit tests ensure that individual components of a system work and return the expected results. In ML systems, unit testing is highly important but needs a different approach because of the stochastic nature of machine learning algorithms. Traditional unit tests in software engineering concentrate on deterministic outputs; however, ML unit tests must handle non-deterministic outputs, probabilistic behavior, and dependency on external data sources.

Why Unit Testing Is Important in ML Models

ML models include several stages, such as data preprocessing, feature engineering, training, and inference. Errors can happen at any of these stages, resulting in poor performance, wrong predictions, or biased outputs. Unit tests help in the early detection of errors in the ML development pipeline. They also ensure reproducibility by confirming that models can produce consistent outputs with the same inputs. They prevent issues caused by incorrect encoding, scaling, or missing values and ensure they are within acceptable ranges and distributions.

Key Components to Test in ML Models

Effective unit testing in ML needs to cover some components. In this section, we will discuss various components that we need to write unit tests for. When I was writing my first unit test for ML, I was very excited. I have written many unit tests for regular deterministic codes, and writing unit tests for stochastic codes was a new experience.

Data Preprocessing and Feature Engineering Tests

We talked in previous chapters about data preprocessing. Data preprocessing is one of the first steps in building machine learning models that prepare and clean the data for the models. This process includes encoding categorial data, handling missed values, and feature scaling. Feature engineering is the art of preparing effective inputs (features) from existing data for improving model training and prediction performance.

Feature transformations are specific mathematical processes that are applied on features. These processes include normalizing and standardizing the features, logarithmic, and other types of transformations. Feature transformations are a well-known source of errors. They often fit into training data but fail when used in production data with outliers. For instance, production data have a new unseen category in specific categorial data. These transformations must be tested to ensure that they behave as expected over various datasets. The following Python code checks if a `MinMaxScaler` transformation scales input values to the range [0,1] correctly. This feature scaling brings the values of all features into a specific range. If we do not scale features, features with large values will dominate the learning algorithm during the training.

Listing 8-1. Testing MinMaxScaler Transformation

```
import pytest
import numpy as np
from sklearn.preprocessing import MinMaxScaler

def test_minmax_scaler():
    scaler = MinMaxScaler()
    X = np.array([[10], [20], [30], [40], [50]])
    X_scaled = scaler.fit_transform(X)

    assert np.all(X_scaled >= 0) and np.all(X_scaled <= 1), "Scaling
    failed: Values are outside [0,1] range"

test_minmax_scaler()
```

In the above code, `pytest` is used. `Pytest` is a well-known library for unit testing in Python.

Model Training and Output Consistency Tests

As ML models rely on stochastic processes, training the same model on the same data should result in approximately similar results. Pay attention that I said approximately. The following code test ensures model reproducibility.

Listing 8-2. Sample Code for Reproducibility Test

```
import tensorflow as tf
import numpy as np

def test_model_reproducibility():
    np.random.seed(42)
    tf.random.set_seed(42)

    model = tf.keras.Sequential([
        tf.keras.layers.Dense(10, activation='relu', input_shape=(5,)),
        tf.keras.layers.Dense(1, activation='sigmoid')
    ])

    model.compile(optimizer='adam', loss='binary_crossentropy')

    X_train = np.random.rand(100, 5)
    y_train = np.random.randint(0, 2, size=100)

    history_1 = model.fit(X_train, y_train, epochs=1, verbose=0)
    history_2 = model.fit(X_train, y_train, epochs=1, verbose=0)

    assert np.isclose(history_1.history['loss'][0], history_2.
    history['loss'][0], atol=0.01), "Training is not
    reproducible"    model.save("saved_model.h5")
    print("Model saved as saved_model.h5")
test_model_reproducibility()
```

Expected Results

```
Model saved as saved_model.h5
```

Inference Tests

At all times, the model's inference function should produce outputs in the expected format. The following code ensures that predictions are in a valid probability range for classification tasks.

Listing 8-3. Model Inference Test

```
import numpy as np
import tensorflow as tf
def test_model_inference():
    np.random.seed(42)
    X_test = np.random.rand(10, 5)  # Simulating test samples

    model = tf.keras.models.load_model("saved_model.h5")  # Load a
    trained model
    predictions = model.predict(X_test)

    assert np.all(predictions >= 0) and np.all(predictions <= 1),
    "Inference outputs are not valid probabilities"
    print("Random test passed - predictions are valid probabilities.")
    print("Random predictions:\n", predictions)

    # ==============================
    # 2. MANUAL NUMERICAL TEST INPUT
    # ==============================
    manual_input = np.array([
        [0.1, 0.2, 0.3, 0.4, 0.5],
        [1.0, 0.0, 0.5, 0.2, 0.1],
        [0.9, 0.9, 0.9, 0.9, 0.9],
    ])

    manual_predictions = model.predict(manual_input)

    # Validate range again
    assert np.all(manual_predictions >= 0) and np.all(manual_predictions
    <= 1), \
        "Manual inference outputs are not valid probabilities"
```

```
    print("\nManual numerical test passed.")
    print("Manual inputs:\n", manual_input)
    print("Manual predictions:\n", manual_predictions)test_model_
    inference()
```

Expected Results

```
1/1 ──────────────────── 0s 59ms/step
Random test passed - predictions are valid probabilities.
Random predictions:
 [[0.55428994]
 [0.4072645 ]
 [0.56010836]
 [0.47687516]
 [0.410865  ]
 [0.4682452 ]
 [0.39856982]
 [0.4200521 ]
 [0.5494172 ]
 [0.4642907 ]]
1/1 ──────────────────── 0s 77ms/step

Manual numerical test passed.
Manual inputs:
 [[0.1 0.2 0.3 0.4 0.5]
 [1.  0.  0.5 0.2 0.1]
 [0.9 0.9 0.9 0.9 0.9]]
Manual predictions:
 [[0.4471692 ]
 [0.40451482]
 [0.4055066 ]]
```

We have two functions in the above code; the first one builds a simple model, and then it saves that to the disk, and the second one loads the model and runs that for sample inputs.

Automating Unit Testing for ML Pipelines

ML pipelines are complex, involving data preprocessing, ingestion, training, evaluation, and deployment. Automating unit tests is needed for continuous validation of each component. Some of the popular frameworks for ML unit testing include

- `pytest`: Standard testing framework for Python-based ML workflows
- `unittest`: Built-in Python testing module for structured test cases
- `TensorFlow Model Analysis (TFMA)`: Is used for validating ML models in TensorFlow environments
- **Great Expectations**: For data integration and validation checks

The following code shows a CI/CD pipeline configuration in **GitHub Actions** that automates ML testing.

Listing 8-4. CI/CD Pipeline Configuration in GitHub Actions

```
name: ML Unit Testing Workflow
on: [push]

jobs:
  test:
    runs-on: ubuntu-latest
    steps:
      - name: Checkout repository
        uses: actions/checkout@v2

      - name: Set up Python
        uses: actions/setup-python@v2
        with:
          python-version: 3.8

      - name: Install dependencies
        run: pip install pytest tensorflow numpy scikit-learn

      - name: Run tests
        run: pytest tests/
```

This configuration is used to automatically run unit tests whenever some changes are pushed to the repository. This approach ensures that any new code modifications would not introduce bugs.

Challenges in Unit Testing for ML Models

ML models show unique challenges in unit testing that are different from traditional software. These challenges include

- **Non-deterministic Outputs:** In normal software, if you give the same input to the tests, the results are always the same, but in ML models, because of the model's random nature, stochastic gradient descent (SGD), and data shuffling, the results are different each time but not necessarily wrong. So, if the model is retrained, the produced results and accuracy would be slightly different.
- **Changing Data Distributions:** A model can pass tests today but may fail in production if data drifts over time. For example, assume we have a fraud detection model and the result is accurate now. But our model is trained using last year's fraud detection pattern, and if fraud tactics change, model accuracy degrades, although nothing has changed in the code.
- **Complex Dependencies:** ML models depend on several external factors, such as hyperparameters, feature engineering steps, and training data. A small change in any of these parameters can affect the finding model prediction. For example, assuming nothing else has changed, if we just change the normalization of the data, the final model prediction will be affected.

Despite these challenges, implementing a structured testing strategy is necessary for mitigating the risks and ensuring consistent model behavior before deploying it to production.

Addressing Model Performance Failures

Machine learning models are not all the times perfect. Even after extensive testing and tuning, performance failures can happen due to a variety of reasons, such as underfitting, overfitting, poor generalization, data drift, and concept drift. Addressing these failures is required for ensuring that models remain fair, accurate, and reliable when deployed in real-world applications.

Diagnosing ML performance failures needs analyzing data, model behavior, and external influences that can affect predictions. In this section, we explore common causes of model performance failures and provide systematic debugging techniques to debug, diagnose, and resolve the failures effectively.

Common Causes of Model Performance Failures

Machine learning models can fail in various ways. We have discussed some cases in ML models that can fail. Understanding the root cause of these failures is important before applying fixes. In this section, we discuss the most well-known reasons for poor model performance.

Overfitting: The Model Memorizes Instead of Generalizing

Overfitting happens when a model learns patterns highly specific to the training data, rather than generalizing to unseen data.[45] If you see high accuracy in training data but poor accuracy on validation/test data or large differences between training and test loss, your model probably suffers from overfitting.

A simple way for detecting overfitting is by comparing training and validation loss curves. The following code shows how to detect overfitting in an example.

Listing 8-5. Detecting Overfitting Using Loss Values

```
import matplotlib.pyplot as plt

# Simulated loss values from model training
epochs = range(1, 21)
train_loss = [0.4, 0.35, 0.30, 0.25, 0.22, 0.20, 0.18, 0.15, 0.14, 0.13,
0.12, 0.11, 0.10, 0.095, 0.09, 0.085, 0.08, 0.075, 0.07, 0.065]
```

```
val_loss = [0.42, 0.38, 0.36, 0.34, 0.35, 0.37, 0.39, 0.40, 0.42, 0.45,
0.47, 0.50, 0.52, 0.55, 0.57, 0.60, 0.62, 0.65, 0.67, 0.70]

plt.plot(epochs, train_loss, 'b-', label="Training Loss")
plt.plot(epochs, val_loss, 'r-', label="Validation Loss")
plt.xlabel("Epochs")
plt.ylabel("Loss")
plt.title("Overfitting Detection")
plt.legend()
plt.show()
```

In the above code, we have used some data that we can plot. Figure 8-1 shows how train loss and validation loss diverge in overfitting and how you can detect overfitting using the chart below.

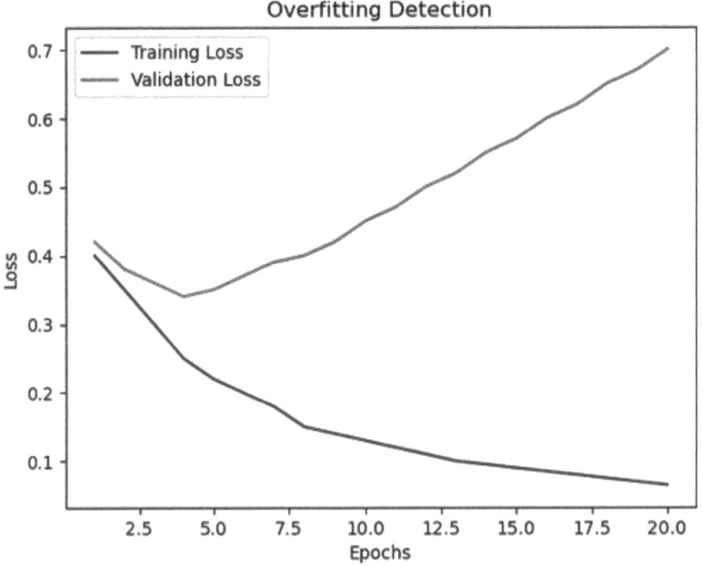

Figure 8-1. *Train Loss and Validation Loss for Overfitting Detection*

If the validation loss increases while the training loss decreases, the model is overfitted.

How to Fix Overfitting

There are some procedures for fixing overfitting. Below are these techniques:

- **Regularization:** Use L1 (Lasso) or L2 (Ridge) regularization to penalize large weights and encourage simplicity.
- **Dropout:** In deep learning, adding some dropout layers can prevent overfitting. Dropouts randomly deactivate neurons during training.
- **Reduce Model Complexity:** A highly complex model with many parameters might not learn patterns and result in a model too customized to the data, which will result in overfitting.
- **More Data and Data Augmentation:** Adding more data to the training dataset is a good way to help the model generalize better.

Underfitting: The Model Is Too Simple

Underfitting happens when a model is too simple to learn a meaningful pattern from the data. Underfitting can be recognized by low accuracy on both validation and training sets. Additionally, when training the model with many epochs does not result in a decrease in training loss, it is another symptom of underfitting. The following code shows underfitting by plotting learning curves.

Listing 8-6. Underfitting by Plotting Learning Curves

```
# Simulated learning curves for underfitting
train_acc = [0.5, 0.52, 0.54, 0.56, 0.58, 0.60, 0.61, 0.62, 0.63, 0.64]
val_acc = [0.48, 0.49, 0.50, 0.51, 0.52, 0.53, 0.54, 0.55, 0.55, 0.56]

plt.plot(range(1, 11), train_acc, 'b-', label="Training Accuracy")
plt.plot(range(1, 11), val_acc, 'r-', label="Validation Accuracy")
plt.xlabel("Epochs")
plt.ylabel("Accuracy")
plt.title("Underfitting Detection")
plt.legend()
plt.show()
```

Similar to the overfitting case, we can have a typical plot for the above code. Figure 8-2 shows this plot.

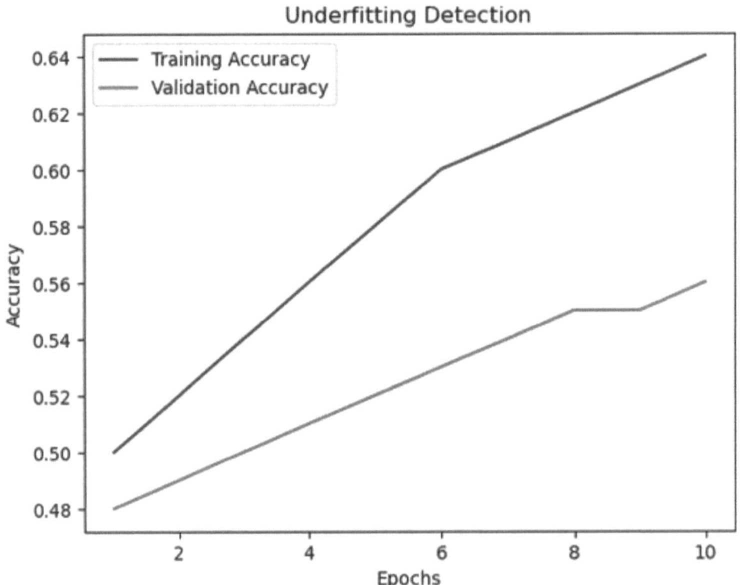

Figure 8-2. *Train Loss and Validation Loss for Underfitting Detection*

How to Fix Underfitting

The first method to get rid of underfitting is to increase model complexity. To make your model more complex, use more neurons, layers, or feature transformations. Also, check your mechanism to prevent overfitting. If regularization is too strong, it can limit the model from learning the required patterns. The last solution is to train for more epochs. Some models in some problems require longer training to converge on a meaningful feature. So, as a summary, if your model is too simple, it may suffer from underfitting, and if it is too complicated, it may suffer from overfitting.

Data Drift and Concept Drift: Changing Data Distributions

We had some talks about data drift and concept drift in different parts of the book. We know that even well-trained models can easily fail in production because of data drift and concept drift. A popular method for detecting drift is comparing the standard deviation and mean of features between production and training datasets. The following code shows how to find data drift by comparing training and production data.

Listing 8-7. Finding Data Drift by Comparing the Train and Production Data

```
import pandas as pd

# Load training and production datasets
train_data = pd.read_csv("train_data.csv")
prod_data = pd.read_csv("production_data.csv")

# Compare feature distributions
for column in train_data.columns:
    train_mean = train_data[column].mean()
    prod_mean = prod_data[column].mean()
    print(f"Feature {column} - Training Mean: {train_mean:.2f}, Production
    Mean: {prod_mean:.2f}")
```

How to Fix Data Drift and Concept Drift

The most famous method of fixing data and concept drift is retraining the model periodically. Regularly updating the model using fresh data is very effective. Also, adaptive learning models (e.g., incremental learning) adjust to changes in data to model in real time. You should be careful that before fixing data and concept drift, you need to distinguish them. So, setting up automated monitoring tools for tracking drift and triggering retraining when necessary is a good practice.

Model Bias and Unfair Predictions

ML models can have biased predictions if they are trained on biased or unbalanced datasets.[46] If you see disproportionate error rates for various groups (e.g., ethnicity, gender) or you see that predictions favor one class over another, your model is probably biased. The following code shows how to check for the biased model.

Listing 8-8. Checking for the Biased Model

```
from sklearn.metrics import accuracy_score

# Sample predictions for two demographic groups
group_A_actual = [1, 0, 1, 1, 0, 1]
group_A_pred = [1, 0, 1, 1, 1, 1]
```

```
group_B_actual = [1, 0, 1, 0, 0, 0]
group_B_pred = [1, 1, 1, 0, 0, 0]

# Calculate accuracy for each group
accuracy_A = accuracy_score(group_A_actual, group_A_pred)
accuracy_B = accuracy_score(group_B_actual, group_B_pred)

print(f"Accuracy for Group A: {accuracy_A:.2f}")
print(f"Accuracy for Group B: {accuracy_B:.2f}")
```

In the above code, a large accuracy gap between different (here, two) groups shows a potential bias. Again, in the above code, I have used some sample data; in your own project, you should replace them with real data.

How to Fix Model Bias

To fix bias, balance the training data. You should ensure fair representation of different groups. Additionally, techniques such as adversarial debiasing or reweighting can be helpful. Also, you can use post hoc bias correction. This correction adjusts predictions for fairness.

Project Title: Testing and Debugging ML Systems

We finish this chapter with a project. Similar to some other chapters, the project of this chapter is a simulation to practice what you have learned in this chapter.

Objective

The purpose of this project is to build a robust debugging and testing framework for an ML pipeline. In this project, you learn how to create unit tests for ML model components and how to identify and debug data pipeline issues. This simulation shows the importance of testing to ensure that ML systems are maintainable and reliable.

Project Description

In this project, you will create a simple ML pipeline that consists of data loading, preprocessing, and model training. Then, you will write some unit tests to check each component, simulate common data pipeline problems, and show debugging practices. We have discussed various issues and how to write a unit test or ML project earlier in this chapter. Like projects of other chapters, this project is organized into modular steps, and each step contains code, explanations, and expected results.

Step 1: Import Required Libraries

Similar to previous projects, in this step, as the first step, we load the required libraries for building the ML pipeline (`NumPy, Pandas, scikit-learn`) and for unit testing (`unittest`). Also, we import `logging` for debugging information during pipeline execution. It is a good practice for you to load all required libraries at the top of your code.

Listing 8-9. Step 1: Import Required Libraries

```
import numpy as np
import pandas as pd
from sklearn.datasets import load_iris
from sklearn.model_selection import train_test_split
from sklearn.preprocessing import StandardScaler
from sklearn.linear_model import LogisticRegression
from sklearn.metrics import accuracy_score

# For unit testing and debugging, we import unittest and logging.
Import unittest
import logging

# Set up basic logging configuration to help in debugging.
Logging.basicConfig(level=logging.INFO, format='%(levelname)s:
%(message)s')
```

Expected Results

This step has no output. It just loads the libraries.

Step 2: Build a Simple ML Pipeline

The code below defines a class for building an ML pipeline. In this code, we define a modular ML pipeline with three classes:

- `DataLoader`: This class loads the Iris dataset. In this book, we have used public datasets, so you can have access to the data without any problem.
- `Preprocessor`: This class scales data using `StandardScaler`.
- `ModelTrainer`: This class trains a logistic regression model and evaluates its accuracy.

The `run_pipeline()` function orchestrates the above steps and prints the final model accuracy. Running this step verifies that the pipeline works perfectly end-to-end.

Listing 8-10. Step 2: Build a Simple ML Pipeline

```
class DataLoader:
    def load_data(self):
        """Load and return the Iris dataset."""
        Data = load_iris()
        X, y = data.data, data.target
        return X, y

class Preprocessor:
    def __init__(self):
        self.scaler = StandardScaler()

    def fit_transform(self, X_train):
        """Fit the scaler on training data and transform it."""
        Return self.scaler.fit_transform(X_train)

    def transform(self, X_test):
        """Transform test data using the fitted scaler."""
        Return self.scaler.transform(X_test)

class ModelTrainer:
    def __init__(self):
        self.model = LogisticRegression(max_iter=200)
```

```
    def train(self, X_train, y_train):
        """Train the logistic regression model."""
        Self.model.fit(X_train, y_train)
        return self.model

    def evaluate(self, X_test, y_test):
        """Evaluate the model using accuracy score."""
        Predictions = self.model.predict(X_test)
        return accuracy_score(y_test, predictions)

# Function to orchestrate the ML pipeline
def run_pipeline():
    # Load data
    loader = DataLoader()
    X, y = loader.load_data()

    # Split the data
    X_train, X_test, y_train, y_test = train_test_split(X, y, test_
    size=0.2, random_state=42)

    # Preprocess the data
    preprocessor = Preprocessor()
    X_train_scaled = preprocessor.fit_transform(X_train)
    X_test_scaled = preprocessor.transform(X_test)

    # Train the model
    trainer = ModelTrainer()
    model = trainer.train(X_train_scaled, y_train)

    # Evaluate the model
    accuracy = trainer.evaluate(X_test_scaled, y_test)
    print(f"Model Accuracy: {accuracy * 100:.2f}%")
    return model, preprocessor, loader

# Run the pipeline once to ensure it works as expected.
Model, preprocessor, loader = run_pipeline()
```

Expected Results

You should see the below output, which shows the accuracy of the model. 100% may look a little weird; however, in the small categorization examples, this accuracy is normal.

```
Model Accuracy: 100.00%
```

Step 3: Unit Testing for ML Pipeline Components

In this step, we use whatever we learned in this chapter. The code below makes a test pipeline. This step creates a test suite using Python's `unittest` framework. The tests created include

- `Data Loader Test`: This test verifies that the data loader returns the correct and expected number of features and samples.
- `Preprocessor Test`: Checks that the preprocessor scales the training data correctly (mean near 0).
- `Model Trainer Test`: This test ensures that the model trains successfully and achieves a minimum accuracy of 70%.

Running `run_unit_tests()` executes all the above tests and prints the results. This ensures that each component of the code functions as expected.

Listing 8-11. Step 3: Unit Testing for ML Pipeline Components

```
class TestMLPipeline(unittest.TestCase):

    def etup(self):
        """Set up the objects for testing."""
        Self.loader = DataLoader()
        self.preprocessor = Preprocessor()
        self.trainer = ModelTrainer()
        # Load data once for all tests.
        Self.X, self.y = self.loader.load_data()
        self.X_train, self.X_test, self.y_train, self.y_test = train_test_
        split(self.X, self.y, test_size=0.2, random_state=42)

    def test_data_loader(self):
        """Test that data loader returns correct shapes."""
```

```
        Self.assertEqual(self.X.shape[0], 150)
        self.assertEqual(self.X.shape[1], 4)

    def test_preprocessor(self):
        """Test that preprocessor correctly scales the data."""
        X_train_scaled = self.preprocessor.fit_transform(self.X_train)
        # Mean of scaled training data should be close to 0.
        Self.assertTrue(np.allclose(np.mean(X_train_scaled, axis=0),
        np.zeros(4), atol=1e-1))

    def test_model_trainer(self):
        """Test that the model can be trained and achieves reasonable
        accuracy."""
        X_train_scaled = self.preprocessor.fit_transform(self.X_train)
        X_test_scaled = self.preprocessor.transform(self.X_test)
        model = self.trainer.train(X_train_scaled, self.y_train)
        accuracy = self.trainer.evaluate(X_test_scaled, self.y_test)
        self.assertGreaterEqual(accuracy, 0.7)

# Running the unit tests
def run_unit_tests():
    suite = unittest.TestLoader().loadTestsFromTestCase(TestMLPipeline)
    unittest.TextTestRunner(verbosity=2).run(suite)

# Execute unit tests
run_unit_tests()
```

Expected Results

```
test_data_loader (__main__.TestMLPipeline.test_data_loader)
Test that data loader returns correct shapes. ... ok
test_model_trainer (__main__.TestMLPipeline.test_model_trainer)
Test that the model can be trained and achieves reasonable accuracy. ... ok
test_preprocessor (__main__.TestMLPipeline.test_preprocessor)
Test that preprocessor correctly scales the data. ... ok

Ran 3 tests in 0.023s

OK
```

The test suite should pass all tests and should return output similar to

```
test_data_loader (__main__.TestMLPipeline) ... ok
test_model_trainer (__main__.TestMLPipeline) ... ok
test_preprocessor (__main__.TestMLPipeline) ... ok
```

Step 4: Debugging Data Pipeline Issues

The code below makes a class that simulates the data pipeline errors. The code of this step simulates a data pipeline error by intentionally removing a feature from the dataset. This causes a mismatch in the number of features and will cause an error during the training stages or preprocessing. After running, the error is caught, and a printed logging message helps you identify the issue. This sample code shows how to use exception handling and logging to debug data pipeline problems.

Listing 8-12. Step 4: Debugging Data Pipeline Issues

```
def simulate_data_pipeline_error():
    print("Simulating data pipeline error: feature dimension mismatch.")
    # Load data correctly first
    X, y = loader.load_data()

    # Introduce error: remove one column from X_train to simulate a
    dimension mismatch
    X_err = X[:, :-1]  # Remove the last feature column
    try:
        # Attempt to split and preprocess the faulty data
        X_train_err, X_test_err, y_train_err, y_test_err = train_test_
        split(X_err, y, test_size=0.2, random_state=42)
        X_train_scaled_err = preprocessor.fit_transform(X_train_err)
        X_test_scaled_err = preprocessor.transform(X_test_err)
        # Train the model with erroneous data
        trainer = ModelTrainer()
        trainer.train(X_train_scaled_err, y_train_err)
    except Exception as e:
        logging.error("Error during pipeline execution: %s", e)
```

```
        print("Debug Tip: Ensure that the input feature dimensions match
        across all pipeline components.")

# Run the error simulation
simulate_data_pipeline_error()
```

Expected Results

```
Simulating data pipeline error: feature dimension mismatch.
```

You should see the above printed message after running the code for this step.

Step 5: Orchestrate Testing and Debugging Workflow

The code below orchestrates the testing and debugging workflow. This code, the `main_testing_debugging()` function, orchestrates the complete workflow. First, it runs the unit tests to ensure that all components are functioning as expected. Then, it simulates the data pipeline error to demonstrate the debugging process.

This final orchestration illustrates how comprehensively designed debugging and testing can show that ML systems are reliable and robust.

Listing 8-13. Step 5: Orchestrate Testing and Debugging Workflow

```
def main_testing_debugging():
    print("=== Starting Testing and Debugging Simulation ===\n")

    # Run unit tests to ensure the pipeline components are working
    correctly.
    Print("Running unit tests...\n")
    run_unit_tests()

    # Simulate a data pipeline error and demonstrate debugging.
    Print("\nSimulating a data pipeline error to demonstrate
    debugging...\n")
    simulate_data_pipeline_error()

    print("\n=== Testing and Debugging Simulation Complete ===")
```

```
# Execute the orchestration if this cell is run directly.
If __name__ == "__main__":
    main_testing_debugging()
```

Expected Results

When you run `main_testing_debugging()`, you will see a detailed unit test output showing that each component works as expected. In addition, you will see the simulated error message and related debugging information and, finally, a final message which indicates that the simulation is complete.

```
Test_data_loader (__main__.TestMLPipeline.test_data_loader)
Test that data loader returns correct shapes. ... ok
test_model_trainer (__main__.TestMLPipeline.test_model_trainer)
Test that the model can be trained and achieves reasonable accuracy. ... ok
test_preprocessor (__main__.TestMLPipeline.test_preprocessor)
Test that preprocessor correctly scales the data. ... ok

Ran 3 tests in 0.022s

OK
=== Starting Testing and Debugging Simulation ===

Running unit tests...

Simulating a data pipeline error to demonstrate debugging...

Simulating data pipeline error: feature dimension mismatch.

=== Testing and Debugging Simulation Complete ===
```

Conclusion

This project shows how to implement debugging and unit testing for an ML pipeline. In this project, we practiced building a modular ML pipeline including data loading, preprocessing, and model training. Additionally, we created unit tests for each component using the `unittest` framework. `Unittest` is one of the most popular libraries for writing unit tests. I have used that in many projects, and its performance is promising. Afterward, we simulated a common data pipeline error logging and also exception handling. Finally, we orchestrated the entire debugging and testing process to ensure the robustness and reliability of ML systems. This project covers all the necessary parts of the chapter about testing and debugging in ML systems. If you want to practice more, I suggest defining various scenarios and writing unit tests for them. You can also use the projects of the other chapters and write unit tests for them.

Summary

Ensuring the robustness and reliability of machine learning systems requires effective debugging and comprehensive testing. In this chapter, we investigated various techniques for finding and resolving common problems in ML workflows. We covered debugging data pipeline issues, unit testing for ML models, and addressing model performance root cause failures with some suggestions for solving the issues.

We started with unit testing for ML models, focusing on the importance of validating individual components, like data preprocessing, feature transformations, model outputs, and inference behavior. We showed real-world approaches for testing feature engineering to ensure model reproducibility and automating testing pipelines using tools such as TensorFlow Model Analysis (TFMA), pytest, and CI/CD integration.

Afterward, we tested data pipeline issues, which are called the source of silent errors in ML workflows. Issues like missing data, data drift, concept drift, incorrect joints, feature leakage, and skewed distributions can considerably impact model performance. We showed debugging techniques, such as data distribution monitoring, statistical tests (Kolmogorov-Smirnov test), and automated data validation for detecting and preventing these issues.

Finally, we discussed model performance failures, which usually arise because of underfitting, overfitting, data drift, concept drift, or biased predictions. We talked about diagnostic methods, including feature distribution analysis, learning curve visualization, and fairness-aware evaluation. Key solutions included model complexity adjustments, regularization, retraining strategies, and fairness-aware ML techniques for improving model generalization and ethical AI deployment.

I am not going to say this chapter is very important, as all chapters of this book provide important information! But by implementing the debugging and testing strategies that we discussed in this chapter, you can build reliable, robust, and fair machine learning models. These practices ensure consistent performance across various practical scenarios. In the next chapter, we will explore explainability and interpretability in ML models and will discuss techniques for improving model accountability and transparency. We are approaching the end of the book, and I start to feel sad about missing such a good friend like you.

CHAPTER 9

Explainability and Interpretability in ML Models

Explainable AI (XAI) is a new field in machine learning (ML) concentrated on developing models in which their decisions could be interpreted and understood by humans. As machine learning models become more complex and neural networks are described as "black boxes," understanding how these models make decisions has become a significant challenge.

In many discussions, *interpretability and explainability* are mentioned together. Although these topics are related, they are not the same. Interpretability addresses how humans will understand the *structure* of the model, for instance, how a decision tree splits or how a linear model assigns weights or how other methods work. On the other hand, explainability concentrates on providing *post hoc explanations* for the model's output. Even for the cases where the model is not transparent (black box models).

Assume that you have developed a project in ML, and you are going to represent that to non-technical stakeholders. Probably the first question, or maybe the second, would be, "How can we trust your model?" If you are confident enough, you can tell them ok, test it, and they will say they will test that, but again, they will ask you to explain how it works, which parameters are included, and what the effect of each is. They may also need this information for the law. Here, explainability will help you. The explainable AI goal is to make the decision-making process of these models transparent and ensure that the logic behind their predictions can be understood, traced, and validated by human stakeholders. This is especially important for meeting regulatory requirements, gaining trust in AI systems, and ensuring that the deployment of AI is fair and ethical.

M. R. Mahdiani, *Mastering Machine Learning Architecture and Solutions*,
https://doi.org/10.1007/979-8-8688-2527-9_9

In this chapter, we talk about explainable artificial intelligence (XAI). We discuss the critical role of XAI in improving trust and transparency in machine learning (ML) systems. Firstly, we show the need for explainability, especially in high-stakes domains like finance and healthcare. In these fields, understanding model decisions is very important. Then, we explore the main challenges in model interpretability and discuss different techniques, such as Shapley Additive Explanations (SHAP) and Local Interpretable Model-agnostic Explanations (LIME), for addressing the challenges. We also talk about model-agnostic and model-specific approaches. Additionally, we examine interpretable ML models such as rule-based systems and decision trees, and we will talk about their advantages and disadvantages. Afterward, we talk about the ethical implications of XAI, highlighting the importance of explainability and regulatory requirements in ensuring accountability and fairness. In this chapter, we will provide numerous practical examples and visual representations to help in understanding XAI and using explainability techniques effectively in your ML applications. So, be ready; there are many interesting and useful topics that we are going to discuss and learn in this chapter.

Importance of Model Explainability

The growing use of machine learning in sensitive applications has increased the need for explainability and transparency. These days, machine learning models are deployed in almost all sectors, including finance, healthcare, and autonomous driving. Some sectors are sensitive, and law enforces them to be explainable. In these cases, decisions made by AI models can directly impact individuals' lives. This makes it necessary to understand how those decisions are made.[47]

Key Reasons for the Need for Explainability

There are many reasons that we should consider explainability in our models. Stakeholders and users need to trust AI models; they need to understand how these models work and which factors affect their decisions. Transparency makes trust and increases the adoption of AI technologies in critical fields. In addition, when AI models make decisions that influence individuals, like medical diagnoses or loan approvals, the logic behind those decisions should be clear. Explainability shows that organizations and model developers are responsible for their AI systems' actions. Additionally, regulations like the General Data Protection Regulation (GDPR) in the European Union require organizations to provide explanations for automated decisions. Explainable AI helps organizations with these legal requirements. Finally, explainability is useful for data scientists for debugging models and improving their performance. By knowing why a model made a certain decision, developers can detect and address biases or weaknesses in the model or the training data.

Key Challenges in Making Models Explainable

Despite the significance of explainability, achieving it is usually difficult, especially for complicated models such as deep neural networks. The "black box" nature of these models means that even the developers who have developed these models cannot fully understand the decision-making process of the models. Before proceeding, think for a moment and try to answer this question: How can we explain black box model predictions? Advanced machine learning models, particularly deep learning architectures, are very complex. These models consist of a huge number of interconnected nodes and layers, in which each node performs complicated calculations. This architecture hinders providing intuitive and clear explanations for the model decisions. Simple models such as decision trees or linear regression are easy to explain and interpret, but may not have the best predictive performance. On the other hand, accurate models such as deep neural networks are usually difficult to interpret, which is a challenge for stakeholders and developers. We must find a way that explain the models but not sacrificing the accuracy and performance. There are different techniques for model interpretability, like SHAP (Shapley Additive Explanations) and LIME (Local Interpretable Model-agnostic Explanations) that can help us. There is no globally accepted method for achieving explainability. Before discussing these techniques, let's talk more about overall models for explainability. The unavailability of a standard model for explainability leads to inconsistencies. The type of required explanation may vary based on the audience. A data scientist might need a technical explanation for debugging the model, while an end user might just need simple logic for a decision. Preparing suitable explanations for various stakeholders is a big challenge. Figure 9-1 shows a sample diagram for the concept of model interpretability. This figure shows the difference between an interpretable model and an opaque "black box" model.

Figure 9-1. *Explainable AI Diagram*

Moving Forward with Explainability

The need for model explainability is becoming more evident as the use of machine learning increases in different industries. Addressing the challenges of standardization, model complexity, and audience-specific explanations is critical for advancing the field of XAI.[48] In this chapter, we explore specific techniques for achieving explainability, such as model-agnostic and model-specific methods. We also talk about the ethical importance of implementing XAI in real-world applications.

Tools and Techniques for Explainability

Explainability methods are required for making machine learning models more interpretable and helping stakeholders understand why and how decisions are made. Depending on complexity and the type of model, various techniques can be used to achieve interpretability. These techniques can be categorized as model-agnostic and model-specific techniques. In the following section, we talk about these categories.

Model-Specific vs. Model-Agnostic Techniques

Based on their compatibility with various types of machine learning models, explainability techniques can be categorized. The two main categories include model-specific techniques and model-agnostic techniques.

Model-Specific Techniques

Model-specific methods are designed for special types of machine learning models. These techniques are tailored to leverage the internal structure of models for providing explanations. Model-specific techniques are usually limited in their application but can provide more detailed insights into the models they are built. Some examples of models for which model-specific techniques are suitable are decision trees and linear regressions. Decision trees are inherently interpretable as their structure shows a series of decisions, based on specific feature splits. You can follow the branches of the tree and easily explain how a prediction is made. Also, linear regression models are interpretable and simple. In linear regression, coefficients directly show the relationship between features and the output. In regression, positive coefficients show a positive impact, while negative ones indicate a negative effect. So, explaining models such as decision trees and regression is not that complicated, and you do not need any specific complicated strategy to explain them. However, not all model explaining is as easy as these.

Decision Trees

Decision trees are one of the most interpretable machine learning models. They represent a series of decision rules that split the data into branches to make predictions based on them. Each internal node of the tree shows a feature, and the branches illustrate conditions applied to that feature, which ultimately lead to the final decision at a leaf node. As decision trees visually map out decision paths, they help stakeholders and data scientists trace step-by-step how a prediction was made.[49] Figure 9-2 shows an example of a decision tree.

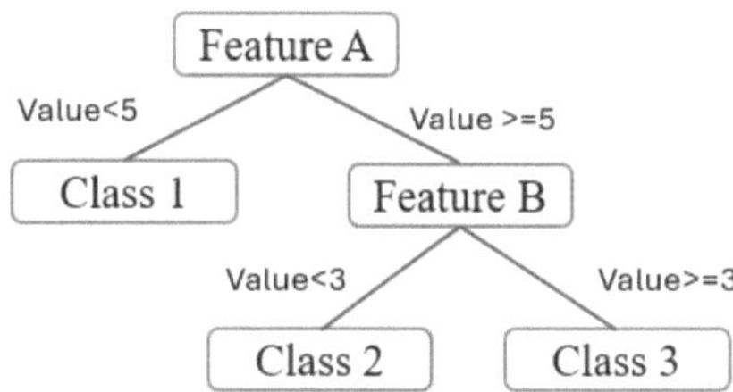

Figure 9-2. *A Typical Decision Tree*

In Figure 9-2, the path taken through the tree to a leaf node can be easily discussed and interpreted. This makes decision trees understandable and transparent for stakeholders.

Rule-Based Models

Rule-based models use a group of "if-then" rules for making predictions. This allows easy interpretability. Rules are fetched from the data and represent relationships between the target outcome and features. For example, a rule might be like this:

```
If Feature A > 3 and Feature B = 'Yes', then predict Class 1
```

It is clear that rules like these, even for non-technical stakeholders, are easy to understand and can be used in justifying the model's decision. Rule-based models are especially useful in domains in which specific conditions determine the outcome. These domains include healthcare and fraud detection.

Model-Agnostic Techniques

Another set of techniques is model-agnostic techniques. These models can be applied to all machine learning models. These methods are more versatile and are widely used while working with black box models such as ensemble methods and deep learning. Permutation feature importance is one of the model-agnostic techniques. This technique measures the effect of each feature by randomly permuting its values and observing how much it affects the model's performance. This method can be used with all models and helps in identifying the important features. Another method is Partial Dependence Plots (PDPs). PDPs hold all features except one or more features constant; then they show the relationship between variable features and the predicted outcome. PDPs offer insights into how a specific feature affects model predictions. Using PDP is one of the easiest ways to find the effect of various parameters on an objective function. I remember when I was in university and students were going to make a machine learning model, I was suggesting to them to have a PDP over the data as early as possible to have some insight about their data and their models. Table 9-1 shows a summary of the differences between model-specific and model-agnostic techniques.

Table 9-1. Comparing Model-Specific and Model-Agnostic

Technique Type	Description	Examples	Advantages	Challenges
Model-Specific	Techniques tailored to specific model types.	Decision Trees, Linear Regression, Logistic Regression, Neural Networks	High interpretability for the given model; efficient due to alignment with model architecture.	Limited to specific models; not transferable across different algorithms.
Model-Agnostic	Applicable to any type of machine learning model.	LIME, SHAP, Permutation Importance, Partial Dependence Plots, ICE (Individual Conditional Expectation)	Universally applicable; can compare results across models; flexible for complex models.	Computationally expensive; requires more processing for large datasets or complex models.

A category of model-agnostic methods is local interpretability methods. These methods provide explanations for each prediction rather than trying to interpret the entire model. These methods help in understanding why a model made a specific prediction for a specific input. Local interpretability is particularly useful when dealing with black box models such as deep learning. I recommend learning these methods well, as they can help explain models to your clients. I have seen that in many companies, using LIME and SHAP is very popular. These techniques are very helpful, and in this section, we talk more about them.

LIME (Local Interpretable Model-Agnostic Explanations)

LIME is a widely used method for local interpretability. LIME's goal is to approximate a complex model with an interpretable, simpler one in the vicinity of the specific input instance. LIME works by perturbing the input data and watching how the model's predictions change. We have talked about perturbing the input of the model before. Do you remember the context? Anyway, this perturbation allows it to fit an interpretable model, like a linear regression, to the local area. This approach gives a straightforward explanation for the factors that mostly influence the prediction. The following sample code shows how LIME can be used for explaining a prediction.

Listing 9-1. LIME for Explaining a Prediction

```
import numpy as np
import sklearn
from sklearn.datasets import load_iris
from sklearn.ensemble import RandomForestClassifier
import lime
import lime.lime_tabular

# Load dataset and train model
iris = load_iris()
X, y = iris.data, iris.target
rf = RandomForestClassifier(n_estimators=100)
rf.fit(X, y)

# Create a LIME explainer
explainer = lime.lime_tabular.LimeTabularExplainer(X, feature_names=iris.
feature_names, class_names=iris.target_names, discretize_continuous=True)

# Explain a prediction
i = 25  # Index of the instance to explain
exp = explainer.explain_instance(X[i], rf.predict_proba, num_features=2)
exp.show_in_notebook()
```

Running this code will generate Figure 9-3.

Figure 9-3. *LIME Analysis Over a Sample of Data*

In the above example, LIME is used to explain why a specific prediction is made by a Random Forest model on the Iris dataset. For this purpose, data is perturbed and fitted to a simpler model. As we mentioned earlier, LIME provides a local explanation to show why the model predicted a value for the specific instance `X[i]`.

SHAP (Shapley Additive Explanations)

Another popular local interpretability method is SHAP. SHAP uses concepts from cooperative game theory to assign contributions to every feature in a prediction. SHAP values help in determining the importance of every feature by quantifying its contribution to the model's output. One of the main advantages of SHAP is that it gives consistent and theoretically grounded explanations. This explanation makes SHAP a strong tool for understanding black box models. In the following, the code demonstrates a simple representation of how SHAP can be used.

Listing 9-2. A Typical Example of SHAP

```
import shap
import xgboost
import numpy as np
import pandas as pd
from sklearn.datasets import fetch_california_housing

# Load dataset and train model
ocus rnia = fetch_california_housing(as_frame=True)
X = ocus rnia.data
y = ocus rnia.target

# Train model
model = xgboost.XGBRegressor().fit(X, y)

# Create a SHAP explainer
explainer = shap.Explainer(model, X)

# Calculate SHAP values for a one-row DataFrame.
Shap_values = explainer(X.iloc[0:1])
# Visualize the explanation for the first (and only) instance.
Shap.plots.waterfall(shap_values[0])
```

This code will result in Figure 9-4.

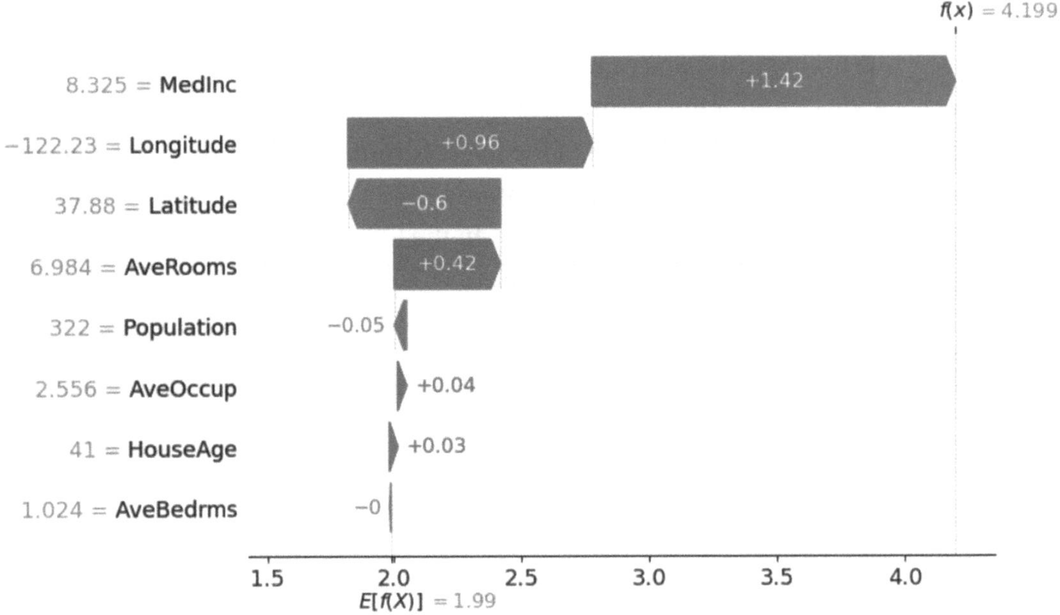

Figure 9-4. *The Result of a Sample SHAP Analysis*

In the above code example, SHAP is used for explaining the predictions that are made by an XGBoost model on the Boston housing dataset. The resulting waterfall plot visualizes the effect of each feature on the model's prediction for a particular instance. Now, after talking about various techniques of explainability, answer this question. How will you decide which method is the best selection for your project? In the next section, we help you answer this question.

Importance of Choosing the Right Explainability Technique

Selecting the correct explainability technique depends on the complexity of the model, the type of model being used, and the intended audience. For simple, interpretable models such as linear regression and model-specific methods are enough. However, for complex black box models, model-agnostic techniques such as SHAP and LIME are usually required for providing insights into the model's behavior. Now, I ask another question.[50] How will you decide to use LIME or SHAP in your project?

Pros and Cons of Using Interpretable Models

Interpretable models offer significant advantages, particularly in terms of accountability and transparency. However, they also have some limitations. In this section, we discuss the main pros and cons of using interpretable machine learning models.

Pros of Interpretable Models

One of the most important benefits of interpretable models is that they offer transparent and clear decision-making logic. This transparency allows stakeholders, including users and regulators, to easily understand why and how a particular prediction is made. It is clear how this transparency can make accountability and trust. Interpretable models build trust, as their predictions can be justified and explained. I think there is no need to explain more about how interpretable models can build trust, and we can talk about the benefits of explainable models in debugging. Understanding how features affect a model's output allows data scientists to detect biases or errors in the model. For instance, in the case where a decision tree gives too much importance to an irrelevant feature. Additionally, in regulated industries, like healthcare or finance, it is required to provide explanations for automated decisions. I said it is required; it means interpretable models are needed to comply with regulations like GDPR, which requires explainability in automated decision-making.

Cons of Interpretable Models

Interpretable models, like linear regression or decision trees, often lack the required complexity to capture complicated relationships in data. While they are providing transparency, they usually underperform in comparison to more complicated black box models. This difference is more sensible when dealing with nonlinear or high-dimensional data. Additionally, when datasets get larger, even decision trees become less interpretable. Complex decision trees with many branches might become challenging to understand, which decreases their transparency. Finally, interpretable models may not always have high predictive accuracy. Complex models like ensemble methods or deep neural networks usually outperform interpretable models in terms of accuracy. This makes interpretable models less accurate and less suitable for some applications. Table 9-2 compares decision trees vs. rule-based models.

Table 9-2. *Decision Trees vs. Rule-Based Models*

Model Type	Description	Strengths	Weaknesses
Decision Trees	Tree-structured models for decision-making	Easy to visualize, transparent	May overfit or become too complex
Rule-Based Models	Use "if-then" rules derived from data	Easy to understand, simple	Limited ability to capture complex relationships

Explainability in the Real World

In this section, we will talk more about real cases in which explainability can be helpful. We start with bias, and then we continue our discussion about other cases where explainability can help.

Explainability in Bias Detection and Mitigation

One of the main reasons that we need explainability is to ensure that the model results are fair. To get that more effectively, we need to talk more about bias. In previous chapters, we talked about bias; we mentioned how to detect it and how to mitigate that. Here, in this chapter, we focus on how XAI techniques help us with opening up black box models and audit cases of unfairness. For this purpose, we use methods such as SHAP and LIME to clarify why some certain decisions of the model are biased. As a reminder, bias in machine learning is an important ethical issue that can result in indiscriminate outcomes. Bias reduces model performance and has unintended social consequences. Bias usually stems from incomplete or biased training data, which ultimately affects the behavior of machine learning models.[32] Let's review some case studies about bias.

In 2018, Amazon had to scrap an AI-based recruitment tool that was later discovered to be biased against women. The tool was trained using historical hiring data. The data primarily reflected male candidates. This resulted in the AI model learning from data to favor resumes that were like those of male applicants in its training data. Amazon tried to fix the model by removing gender-related keywords. But the historical bias was deeply embedded in the data, which was hardening the problem. This case shows the significance of fair data collection and preprocessing techniques to get rid of bias before training the model. So there are two points that we can learn from this case study:

the first is do not include extra data in your training data. For example, for resumes in training data that are all for men, why should you specify the gender? This field can result in bias. Another point is that even if we remove this field, there is no guarantee that your model would be fair. Also, another lesson that we can learn from this case study is the importance of explainability, which helps to detect bias.

Another example is about the COMPAS algorithm. The COMPAS algorithm was used in the US judicial system for predicting the probability of defendants reoffending. It was discovered that COMPAS systematically predicted a higher recidivism risk for African American defendants compared to Caucasian defendants. Audits by researchers showed disparities in predictions. Tools such as LIME and SHAP were used for understanding the decision-making process of the model. The findings resulted in discussions about applying equal opportunity metrics for balancing the true positive rates over various racial groups. Figure 9-5 shows the stages in which bias can be injected into the machine learning pipeline. These stages include data collection, preprocessing, model training, and deployment. This figure also shows the corresponding mitigation techniques, like data augmentation, diverse sampling, fairness-aware algorithms, and post-deployment audits.

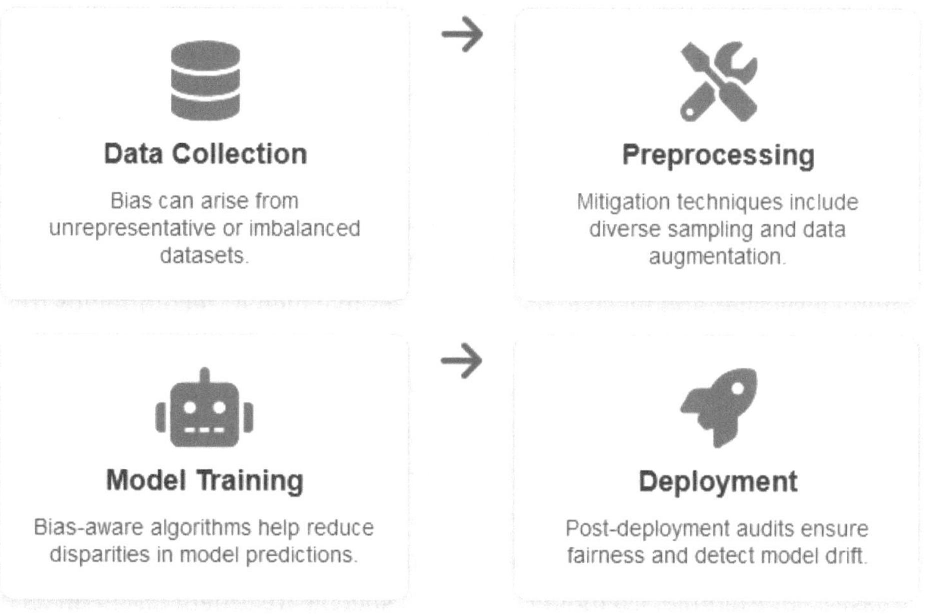

Figure 9-5. *Overview of Bias, Introduction, and Mitigation Techniques in Machine Learning*

In the above example, we investigate how explainability can help us to get rid of bias.

Use Cases and Ethical Importance of XAI

Explainable AI (XAI) is getting increasingly critical in machine learning applications, specifically in industries in which decision-making transparency is important. Using machine learning in high-stakes domains such as finance and healthcare requires that the models be both accurate and explainable. Understanding the significance of explainability in these fields is important for ensuring ethical deployment and also compliance with regulatory requirements.

Explainability in Healthcare and Finance

In this section, we will investigate the role of explainability in sensitive industries such as healthcare and finance.

Healthcare

In healthcare, the significance of explainability cannot be overstated. Machine learning models are being used to diagnose diseases, predict patient outcomes, recommend treatments, and even help in surgical procedures. However, it is required for healthcare professionals to be aware of the logic behind these model predictions and ensure proper care and patient safety.[51] AI models predict the probability of certain diseases based on patient data. Doctors need to be aware of how the model reached this diagnosis to validate its recommendations. For instance, a model predicting heart disease risk, based on patient data, should be able to explain the features that lead to this prediction, such as cholesterol level, age, and lifestyle habits. In cases in which AI recommends treatment plans, physicians need to explain why a specific treatment is being suggested. This is particularly critical when patients and their families want to understand the reason behind decisions. It is their right to ask for the reasons, and telling them AI made this decision is not satisfying for them. Lack of explainability in healthcare can result in potentially life-threatening consequences. We know that AI can make mistakes, it can be attacked, and there are several other reasons that we cannot trust AI decisions in sensitive cases. For example, if a model suggests a specific treatment based on its trained data, healthcare providers need to verify it. Verification helps in mitigating the risk of the suggestion by understanding the underlying reasons for the recommendation.

Finance

In the financial industry, explainability is critical for loan approvals, fraud detection, risk assessment, and investment decision-making. A mistake in these fields can have terrible consequences. Financial institutions are subject to stringent regulations. For them, decisions made by AI models must be justifiable and transparent to both clients and regulators.

AI models are frequently used to handle loan applications and determine creditworthiness. However, it is the right of applicants to know why their application was denied or approved. Explainable models help financial institutions in providing reasons for their decisions, avoiding discrimination, and ensuring fairness. In addition to explainability, you should ensure that your models are not biased. We talked about detecting and mitigating bias a while ago. You should be aware that if your model makes a biased decision, for example, not approve a loan because of the race of the applicant, you may be sued. AI models are frequently used in fraud detection. They flag potentially fraudulent transactions based on analyzing patterns in customer behavior. Explainability helps analysts to understand why a specific transaction was flagged as fraudulent. That allows them to avoid false positives and take appropriate actions. Also, machine learning models are used in providing investment recommendations. Clients need to understand the reasons behind investment decisions, particularly in situations with high risk. Transparent models build trust with clients and comply with regulations.

Regulatory Requirements for Explainability

With the increasing use of machine learning in high-stakes industries, regulatory bodies have introduced guidelines that require explainability and transparency in AI models. Compliance with these regulations is required to ensure the ethical deployment of AI systems which protects the rights of individuals. The European Union's GDPR emphasizes that it is the right of individuals to receive an explanation about decisions that affect them and are made by automated systems. It makes sense that no one can escape the law by just saying this decision is made by AI. Article 22 of the GDPR is about protecting individuals from automated decisions without human intervention. This has considerable implications for machine learning models, especially in healthcare and finance, where explainability is necessary to comply with these regulations.

The purpose of the proposed AI Act in the European Union is to regulate the use of AI systems based on the level of their risk. High-risk AI systems (definitely you know I mean the ones in finance and healthcare or any other sensitive industry) have to meet more transparency requirements. This includes ensuring that users understand the limitations and capabilities of the AI system. Because of these limitations, users need to provide clear explanations of how decisions are made. We have similar regulations in the United States. In the United States, the ECOA mandates that individuals must not be discriminated against when applying for credit. Additionally, when an applicant is denied, they are entitled to an explanation about the reason for being denied, and clearly, you cannot tell them AI made this decision; ask AI for the explanation. For example, if a loan application is denied, the applicant has the right to ask for the reasons behind the decision. This regulation shows the need for explainable models in finance to avoid discriminatory practices and ensure fair treatment of all applicants. Assume that your model denies a person and you do not care about explainability, but after further research, you find that the data that is used for training the model was biased and that the applicant was rejected based on ethnicity. Additionally, the U.S. Food and Drug Administration (FDA) has issued guidelines about the use of AI in healthcare, especially for AI-driven medical devices. These guidelines state that AI models should be transparent and enable healthcare providers to make informed decisions while ensuring patient safety.

Ethical Importance of Explainability

Explainable AI is not only a regulatory requirement but also an ethical necessity. The ethical importance of explainability is due to its ability to reduce bias, empower users, and foster trust in AI systems. We are going to make a better world using AI, so we should be careful that our models are transparent enough and do not make life difficult for some people because of their biased decisions. When users understand the logic behind AI decisions, they can trust more, make informed choices, and hold developers accountable. Explainable AI is required to identify and reduce biases in machine learning models. It reveals which features are contributing to predictions. This transparency helps developers detect and correct biases that could lead to unfair treatment of individuals or groups. Trust is the main component of AI adoption, especially in sensitive industries. When stakeholders, regulators, and end users can trust and understand the decisions made by AI systems, they are more likely to rely on these technologies. This helps these technologies to be more useful and will be used more efficiently. In addition, explainable AI puts humans at the center of decision-making processes. At least at the current level of AI, we need that. We know AI can make mistakes, and in sensitive cases, these mistakes can put human life in danger. By providing clear and understandable explanations, XAI helps human oversight and ensures that AI systems complement rather than replace human judgment. Figure 9-6 shows a visual representation of the importance of explainable AI in finance and healthcare. This figure illustrates decision points, the flow of data, and the role of human oversight.

***Figure 9-6.** Representation of the Importance of Explainable AI in Healthcare and Finance*

Project Title: Explainability and Interpretability in ML Models

In the project of this chapter, we will practice explainability and interpretability in a real-world problem.

Objective

This project shows how to build and explain a machine learning model using both model-agnostic explainability techniques and inherently interpretable models. By doing this project, you will learn to

- Train a decision tree classifier, a model known for its interpretability.
- Visualize the decision rules of the trained model.
- Use SHAP (Shapley Additive Explanations) for providing local and global explanations of model predictions.
- Understand how to handle feature importance and discuss bias/ fairness considerations.

Project Description

Here, we develop a project, and then we use explainability techniques to test our learning from this chapter.

Step 1: Import Required Libraries

The code below imports the required libraries.

Listing 9-3. Step 1: Import Required Libraries

```
import numpy as np
import pandas as pd
import matplotlib.pyplot as plt

from sklearn.datasets import load_iris
from sklearn.model_selection import train_test_split
from sklearn.tree import DecisionTreeClassifier, plot_tree
from sklearn.metrics import accuracy_score

# Import SHAP for model explainability
import shap

# Optional: Set a random seed for reproducibility.
Np.random.seed(42)
```

Explanation

Similar to all other projects in the book, in the first step, we load the libraries required for our project.

- NumPy and Pandas: For data manipulation and numerical operations
- scikit-learn: For loading the dataset, splitting data, training a decision tree classifier, and evaluating its performance
- matplotlib: For visualizing the decision tree and SHAP summary plots
- SHAP: For computing and visualizing feature contributions to predictions, enhancing interpretability

Expected Results

The step should run without any output; if you see outputs, it means something went wrong, and you are in trouble!

Step 2: Load and Prepare the Data

The code below loads and prepares the data for the next steps.

Listing 9-4. Step 2: Load and Prepare the Data

```
iris = load_iris()
X = iris.data
y = iris.target

# For simplicity, we will convert it into a binary classification problem
# (e.g., class 0 vs. classes 1 and 2).
Y_binary = (y == 0).astype(int)

# Split data into training and testing sets
X_train, X_test, y_train, y_test = train_test_split(X, y_binary, test_
size=0.2, random_state=42)

print("Data loaded and split:")
print(f"Training samples: {X_train.shape[0]}, Testing samples: {X_test.
shape[0]}")
```

Explanation

In this step, we load the Iris dataset and convert it to a binary classification problem. This conversion is required for clarity in explanation. The dataset is then split into training and testing.

Expected Results

```
Data loaded and split:
Training samples: 120, Testing samples: 30
```

The output shows a message illustrating the number of training and testing samples.

Step 3: Train a Decision Tree Classifier

The code below trains the decision tree. I have set some default hyperparameters for that, but you can change them.

Listing 9-5. Step 3: Train a Decision Tree Classifier

```
clf = DecisionTreeClassifier(max_depth=3, random_state=42)
clf.fit(X_train, y_train)

# Evaluate the model on test data
y_pred = clf.predict(X_test)
accuracy = accuracy_score(y_test, y_pred)
print(f"Decision Tree Model Accuracy: {accuracy * 100:.2f}%")
```

Explanation

In this step, we instantiate a decision tree classifier with a maximum depth of 3. We select 3 for maintaining interpretability and simplicity. However, you can change the maximum depth and play with it to solidify your understanding. Then, the model is trained on the training set and afterward is evaluated on the test set. Finally, its accuracy is printed to the console.

Expected Results

```
Decision Tree Model Accuracy: 100.00%
```

The output shows the accuracy of the model.

Step 4: Visualize the Decision Tree

The code below plots some visualizations for the decision tree.

Listing 9-6. Step 4: Visualize the Decision Tree

```
plt.figure(figsize=(12, 8))
plot_tree(clf, feature_names=iris.feature_names, class_names=["Not Class
0", "Class 0"], filled=True)
plt.title("Decision Tree Visualization")
plt.show()
```

Explanation

This step visualizes the trained decision tree, illustrating decision rules and how features are used to split the data and lead to a decision. Each node in the tree provides interpretable conditions for classification decisions.

Expected Results

In this step, a figure (Figure 9-7) will be generated, showing the structure of the decision tree. In that, nodes indicate class distributions, feature thresholds, and impurity levels.

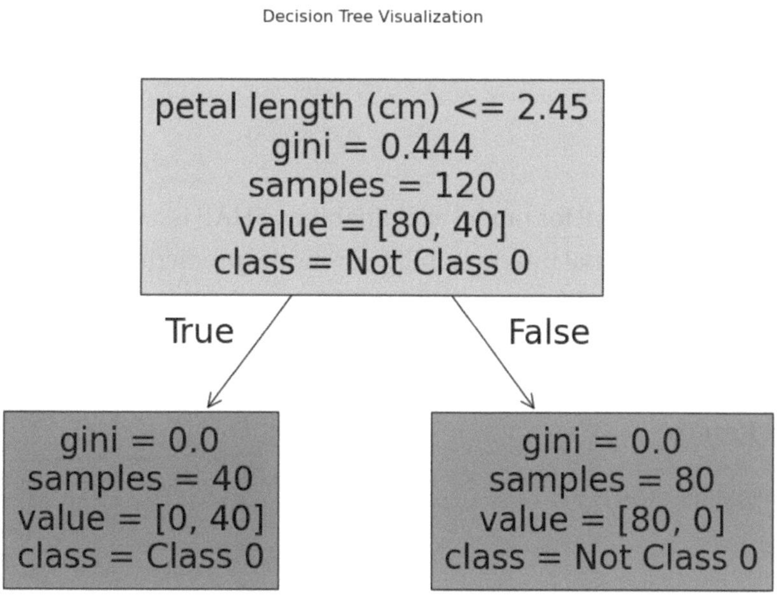

Figure 9-7. *Classification Tree of the Project*

Step 5: Explain Model Predictions with SHAP

The code below shows how to use SHAP to explain the model.

Listing 9-7. Step 5: Explain Model Predictions with SHAP

```
# Create a SHAP TreeExplainer using the trained decision tree
explainer = shap.TreeExplainer(clf)
# Compute SHAP values for the test set
shap_values = explainer.shap_values(X_test)

# Plot a summary plot of SHAP values for the positive class (class 0)
explainer = shap.Explainer(clf, X_train)
shap_values = explainer(X_test)

# Now, shap_values.values should have shape (n_samples, n_features)
print("SHAP values shape:", shap_values.values.shape)  # Expect (30, 4)

# Plot the summary plot using the new SHAP values
shap.summary_plot(shap_values.values, X_test, feature_names=np.array(iris.
feature_names))
```

Explanation

In this step, we use SHAP for model explainability. SHAP is a strong tool that provides both local and global interpretability for model predictions. In this step, we use a `TreeExplainer` tailored for tree-based models to calculate SHAP values for each feature on the test set. The summary plot (Figure 9-8) illustrates the effect of features on the prediction for class 0 (our positive class in the binary setup).

Expected Results

```
SHAP values shape: (30, 4, 2)
```

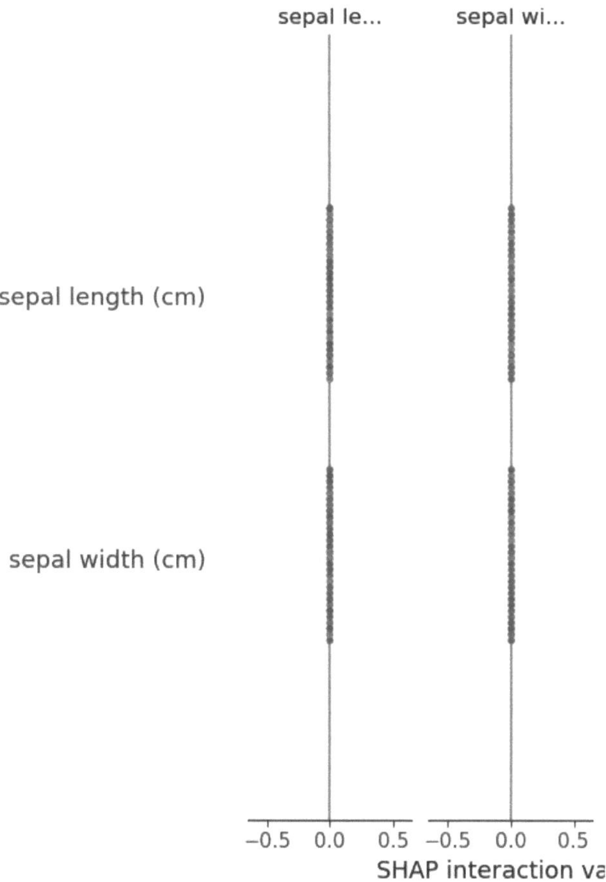

Figure 9-8. *SHAP Summary Plot for Feature Importance*

After running the code, a SHAP summary plot will appear, which shows the distribution of SHAP values over features and illustrates their impact on model output.

Step 6: Addressing Bias and Fairness Considerations

The code below can be used to address bias and fairness.

Listing 9-8. Step 6: Addressing Bias and Fairness Considerations

```
mean_abs_shap = np.abs(shap_values[1]).mean(axis=0)
feature_importance = dict(zip(iris.feature_names, mean_abs_shap))
print("Mean Absolute SHAP Values (Feature Importance):")
for feature, importance in feature_importance.items():
    print(f"{feature}: {importance:.3f}")

# In practice, these values can be compared with domain knowledge to ensure
fair model behavior.
```

Explanation

This code calculates the absolute SHAP values for each feature. This serves as an indicator that shows the overall feature importance. Such analysis can help in identifying potential biases in how features affect predictions. In practical applications, comparing domain knowledge with these insights is crucial for ensuring fairness.

Expected Results

```
Mean Absolute SHAP Values (Feature Importance):
sepal length (cm): 0.167
sepal width (cm): 0.167
```

Conclusion

In this project, we showed the complete workflow for interpretability and explainability in machine learning models. First, we trained an inherently interpretable decision tree classifier on a modified Iris dataset. Following, we used SHAP to explain both local and global model behaviors. We also talked about handling bias and ensuring fairness by analyzing feature importance. It is an important step, and try to learn and use that in your projects. You would definitely need it in serious real-world projects. In each step, we provided clear outputs and visualizations. In this project, our purpose is to make it possible for you to test and validate the interpretability of ML systems by yourself. By doing this project, you should solidify your learning in understanding the importance of model interpretability.

- Visualizing decision rules in interpretable models.
- Applying SHAP for gaining insights into feature contributions. I suggest you try the SHAP part again by yourself without reading the content of the book.
- Discussing potential bias and fairness issues based on explainability outputs.

If you are interested, you can add some extra steps to this project. For example, you can explain the model using LIME.

Summary

The purpose of explainability is to make machine learning models more understandable and transparent for stakeholders. Methods like SHAP and LIME provide insights into the decision-making process of complex models and help regulators, data scientists, and users understand how models make some predictions. These tools break down the contribution of each feature to a model's output and allow stakeholders to have a clear understanding of why a model behaves in a specific way. We mentioned that there are some other reasons for explainability, too. We said some regulations force us to explain the model; it is the right of people who are affected by the model to see how decisions are made, and we need to ensure the decisions that are made by the model are fair, reliable, and trustworthy. Using interpretable models in crucial applications, like finance and healthcare, ensures that decisions are justifiable, fair, and free from unintended biases. This transparency is particularly important in scenarios in which AI-driven decisions have a significant impact on human life, like medical diagnoses or loan approvals. We cannot convince people by telling them that AI made a mistake in diagnosing their disease or that AI denied their loan. The increasing need for explainability is driven by ethical concerns, and regulatory requirements make explainability an essential aspect of responsible AI deployment. Explainability not only improves model performance by facilitating debugging but also helps in regulatory compliance. Regulators will not sue AI; they will sue humans who have used AI without verifying the AI decision. Moreover, enhancing model interpretability builds trust and enables end users to rely on AI systems confidently. This manner ensures models align well with societal expectations and ethical standards.

We finish the chapter here. In the next chapter, we will cover some new trends in various aspects of ML, such as NLP and computer vision, and we will finally finish the book.

CHAPTER 10

Future Trends and Challenges in Machine Learning Systems

In this chapter, we explore modern emerging trends in machine learning (ML) architecture. We will cover the main advancements like Vision Transformers (ViTs), transformer models beyond natural language processing, the rise of agentic AI, and their applications in multi-modal learning. We also include practical samples such as transformer-based image classification. Additionally, we talk about pre-training, self-supervised learning, and fine-tuning, showcasing their potential for revolutionizing ML workflows. Afterward, we talk about AI in edge computing. Edge devices include your personal computers, cellphones, and IoT devices, and for many reasons, running ML models on them is interesting. By that, users have higher privacy, especially in sensitive data; they will experience less delay and can use their models in places with limited network coverage. However, running models in an edge device is not easy because of the limited resources of the edge device. In this chapter, we will discuss various benefits and challenges of ML in edge devices, and we will provide some suggestions to overcome challenges.

Transformer Models

The world of machine learning is observing a quick advancement in model architectures. This leads to a new age of efficient and powerful systems. Staying updated with these trends is critical for developers, researchers, and organizations targeting to

M. R. Mahdiani, *Mastering Machine Learning Architecture and Solutions*,
https://doi.org/10.1007/979-8-8688-2527-9_10

harness state-of-the-art methods for their applications.[52] In this section, we discuss the coming trends in ML architectures, such as transformer models. Additionally, we talk about how these technologies are changing the future of machine learning.

Transformer Models Beyond NLP

Transformer architectures are initially designed for natural language processing. Now they are becoming the main force behind many ML applications. This is because of the transformer's ability to capture complex relationships in data. Transformer-based models, like GPT (Generative Pre-trained Transformer) and BERT (Bidirectional Encoder Representations from Transformers), have shown their effectiveness not only in text but also in other domains, such as reinforcement learning and computer vision.[53] The code below shows a sample code for defining a transformer model.

Listing 10-1. Sample Code for the Transformer Model

```
import torch
from transformers import BertTokenizer, BertForSequenceClassification

# Define a function to use BERT for text classification
def classify_text(text, model, tokenizer, max_length=128):
    # Tokenize input text
    inputs = tokenizer(
        text,
        max_length=max_length,
        padding='max_length',
        truncation=True,
        return_tensors='pt'  # PyTorch tensors
    )

    # Get model predictions
    with torch.no_grad():  # No gradient computation for inference
        outputs = model(
            input_ids=inputs['input_ids'],
            attention_mask=inputs['attention_mask']
        )
```

```
    # Get predicted class (0 or 1 for binary classification)
    logits = outputs.logits
    predicted_class = torch.argmax(logits, dim=1).item()

    return predicted_class

# Example usage
def main():
    # Load pre-trained BERT model and tokenizer
    model_name = 'bert-base-uncased'
    tokenizer = BertTokenizer.from_pretrained(model_name)
    model = BertForSequenceClassification.from_pretrained(model_name,
    num_labels=2)

    # Sample text
    sample_text = "Transformers are revolutionizing machine learning!"

    # Classify the text
    prediction = classify_text(sample_text, model, tokenizer)

    # Print results
    print(f"Input text: {sample_text}")
    print(f"Predicted class: {prediction} (0 = negative, 1 = positive)")

# Run the example
if __name__ == "__main__":
    main()
```

The result should be like this.

```
Input text: Transformers are revolutionizing machine learning!
Predicted class: 1 (0 = negative, 1 = positive)
```

Vision Transformers (ViTs)

The transformer architecture has been adapted for computer vision in the shape of Vision Transformers (ViTs). ViTs use the self-attention mechanism for processing image patches. Unlike traditional convolutional neural networks (CNNs), ViTs do not rely on convolutions for extracting features. Instead, they use self-attention for learning relationships between image patches. This change has led to considerable performance improvements in image classification tasks.[54] The code below shows a sample code of using vision transformer models.

Listing 10-2. Sample Code for Vision Transformer Models

```
import torch
import torch.nn as nn

# Define a minimal Vision Transformer
class MiniViT(nn.Module):
    def __init__(self, image_size=32, patch_size=4, num_classes=10, dim=64,
    num_heads=4):
        super(MiniViT, self).__init__()
        num_patches = (image_size // patch_size) ** 2
        patch_dim = 3 * patch_size * patch_size

        self.patch_embed = nn.Linear(patch_dim, dim)
        self.cls_token = nn.Parameter(torch.randn(1, 1, dim))
        self.pos_embed = nn.Parameter(torch.randn(1, num_patches + 1, dim))
        self.attention = nn.MultiheadAttention(dim, num_heads)
        self.classifier = nn.Linear(dim, num_classes)

    def forward(self, x):
        batch_size = x.size(0)
        x = x.unfold(2, 4, 4).unfold(3, 4, 4).reshape(batch_size, -1, 48)
        x = self.patch_embed(x)
        cls_tokens = self.cls_token.expand(batch_size, -1, -1)
        x = torch.cat((cls_tokens, x), dim=1) + self.pos_embed
        x, _ = self.attention(x, x, x)
        return self.classifier(x[:, 0])

# Example usage
x = torch.randn(2, 3, 32, 32)  # Random 32x32 RGB images
model = MiniViT()
output = model(x)
print(f"Input shape: {x.shape}, Output shape: {output.shape}")
```

We expect a result like this.

```
Input shape: torch.Size([2, 3, 32, 32]), Output shape: torch.Size([2, 10])
```

Applications of Transformers in Multi-modal Learning

Transformers are also useful in **multi-modal learning**, where data from various modalities (e.g., image, text, video) is combined. Models such as CLIP by OpenAI use transformers for understanding both images and their textual descriptions, which leads to interesting results in tasks like visual question answering or image captioning. This trend suggests a future in which ML models can interact with the world and understand in a more humanlike and integrated way.

The following is a sample code that shows how to implement a Vision Transformer using the `transformers` library in Python.

Listing 10-3. Implementing a Vision Transformer Using `transformers`

```
from transformers import ViTForImageClassification, ViTFeatureExtractor
from PIL import Image
import requests

# Load pre-trained Vision Transformer model
model = ViTForImageClassification.from_pretrained('google/vit-base-
patch16-224')
feature_extractor = ViTFeatureExtractor.from_pretrained('google/vit-base-
patch16-224')

# Load and preprocess the image
url = 'https://example.com/image.jpg'
image = Image.open(requests.get(url, stream=True).raw)
inputs = feature_extractor(images=image, return_tensors="pt")

# Make a prediction
outputs = model(**inputs)
logits = outputs.logits
predicted_class_idx = logits.argmax(-1).item()
print("Predicted class index:", predicted_class_idx)
```

This code shows the application of a Vision Transformer model for classifying images. It demonstrates how transformer-based models are now being leveraged across various domains. There are some other emerging technologies, such as self-supervised learning and federated learning. We have discussed them in Chapter 1. Here, we will not repeat ourselves. If you do not remember them, you can go back to Chapter 1.

Agentic AI and Autonomous Systems

In addition to the basic models such as transformers, a critical trend is **Agentic AI**. This trend concentrates on making systems that can plan, reason, and remember. In addition, agents can call different tools to do more complex tasks. Agents are capable of using a reasoning engine (usually a large language model) to interact with the environment. They can break big, complicated tasks into some smaller steps and define a reasonable workflow for doing those tasks. Agents typically have several characteristics including

- **Planning**: They can advise several steps to reach a defined goal.
- **Memory**: They can remember previous acts and can have short-term and long-term memory.
- **Tool Use**: Agents are capable of calling different APIs, running code, querying databases, and interacting with the real world using defined devices.
- **Self-Correction**: Agents can learn from their mistakes. They can see the results and improve their act.

Agentic AI needs a robust infrastructure and a strong MLOps team. In addition, they can be combined with edge devices to provide a more powerful system. In the next step, we talk more about edge computing.

Edge Computing

IoT and edge computing have seen considerable growth in recent years. This growth has increased the number of connected devices and has raised the requirement for real-time data processing. Before, data generated by IoT devices was sent to a centralized server for analysis. This approach usually introduces privacy concerns and latency. However, today, with advancements in AI and edge computing, it is possible to process data closer to the source or at the "edge" of the network.

Some examples of edge devices include cameras for facial recognition, vibration sensors, microphones for voice recognition, and autonomous vehicles with real-time video processing. It is clear that these devices can introduce privacy issues. However, today, by processing data close to the source (using edge computing), this issue can be mitigated.

Integration of AI with edge computing in IoT reduces dependency on cloud infrastructure, allows for real-time data processing, and improves privacy by reducing data transmission. This is especially important in applications that need fast decision-making, like smart cities, autonomous vehicles, and healthcare monitoring systems.

The integration of IoT, edge computing, and AI has changed how data is used and analyzed. Edge computing creates new opportunities for scalability, efficiency, and innovation in various industries. In this section, we provide an overview of how AI technologies are being integrated into IoT systems and edge, including the challenges, benefits, and emerging solutions that are transforming industries. To give you some insight into how edge computing works, let's check the sample code below. As simulating the constraint of the resources on edge devices is not possible here, I have provided some Python classes that show the concept of the workflow of a typical edge device. This includes processing the data locally and optionally sending them to the server.

In this code, we have the `EdgeDevice` class that includes the main components including local data preprocessing, which gives the capability of running a local model (Model class) for fast and secure inference; then, it sends aggregated results to a main remote server. The whole workflow is conceptual. It assumes a low-power environment for deployment. This environment can be a single-board computer, like a Raspberry Pi or an NVIDIA Jetson Nano.

Listing 10-4. Edge AI Deployment Pipeline

```
class EdgeDevice:
    def __init__(self, model):
        self.model = model

    def process_data(self, data):
        # Preprocess data (e.g., sensor readings)
        processed_data = self.preprocess(data)
```

```
        # Run inference using the local model
        prediction = self.model.predict(processed_data)  # Replace with
        actual inference

        # Optionally, send results or updates to a central server
        self.send_to_server(prediction)

        return prediction

    def preprocess(self, data):
        # Implement data preprocessing specific to the edge device
        # For this example, we'll just return the data as is
        return data

    def send_to_server(self, data):
        # Implement logic to send data to a server (e.g., for aggregation)
        print(f"Data sent to server: {data}")

class Model: # A very simple model for the example
    def predict(self, data):
        # Simulate prediction (replace with a real model's prediction)
        return f"Prediction for {data}: {random.random()}"

# Example Usage
my_model = Model() # Replace with your actual model
edge_device = EdgeDevice(my_model)

sensor_data = {"temperature": 25, "humidity": 60}
prediction = edge_device.process_data(sensor_data)
print(f"Edge Device Prediction: {prediction}")

sensor_data_2 = {"temperature": 30, "humidity": 70}
prediction_2 = edge_device.process_data(sensor_data_2)
print(f"Edge Device Prediction 2: {prediction_2}")
```

Expected Results

```
Data sent to server: Prediction for {'temperature': 25, 'humidity': 60}:
0.35409250713083973
```

```
Edge Device Prediction: Prediction for {'temperature': 25, 'humidity': 60}:
0.35409250713083973
Data sent to server: Prediction for {'temperature': 30, 'humidity': 70}:
0.6486073719904244
Edge Device Prediction 2: Prediction for {'temperature': 30, 'humidity':
70}: 0.6486073719904244
```

Benefits of AI at the Edge

Reduced Latency

One of the main benefits of deploying AI at the edge is latency reduction. By processing data locally on edge nodes or IoT devices, it is possible to prevent the delays related to transmitting data to a centralized cloud server. This is especially important in time-sensitive applications, like healthcare monitoring, industrial automation, and autonomous vehicles, where a fraction of a second is critical.

For instance, in autonomous vehicles, edge AI makes real-time decision-making possible. It is doable by processing sensor data directly on the vehicle. This decreases the latency and ensures the vehicle can respond promptly to changes in its environment, including recognizing an obstacle or identifying a pedestrian crossing the road.

Enhanced Privacy and Security

Another main benefit of edge computing is data privacy. Data privacy is a major concern when dealing with sensitive information, like personal user data or medical records. By local data processing at the edge, the need for transmitting sensitive information to a centralized cloud server decreases. This ensures greater privacy for end users and minimizes the risk of data breaches. A good example is in healthcare, where wearable devices are used. These devices are equipped with edge AI that can analyze physiological signals locally and provide health insights without sending raw data to the cloud. This not only preserves user privacy but also decreases the risk of unauthorized access to personal information and data leaks.

Lower Bandwidth Consumption

AI at the edge helps in decreasing bandwidth consumption. This is done by reducing the amount of data that needs to be transmitted to the cloud for processing. Instead of transmitting all raw data, edge devices can process the data locally and only send the relevant insights or results to the cloud. For instance, a smart security camera that is equipped with edge AI can analyze video feeds locally to recognize faces or identify motion. Obviously, this is much more efficient than streaming the entire video feed to the cloud. In the above scenario, the camera only sends relevant footage or alerts, which considerably decreases cloud storage costs and bandwidth usage.

Scalability

Edge AI enables more scalable solutions, especially for large IoT networks. In these networks, centralized processing is impractical because of bandwidth limitations. By distributing computational loads over edge devices, systems can achieve better reliability and scalability. In smart cities, edge computing allows thousands of IoT devices, including surveillance cameras, traffic sensors, and air quality monitors, to operate autonomously and make local decisions. This distributed approach enables cities to scale their smart infrastructure without experiencing network bottlenecks or overloading centralized servers.

Challenges of AI at the Edge

We mentioned that AI at the edge has a lot of benefits from a privacy and latency point of view. However, there are some challenges related to AI at the edge, too, which we discuss in this section.

Limited Computational Resources

One of the key challenges in deploying AI at the edge devices is the limited computational power of edge devices. Despite cloud servers, edge devices like microcontrollers and sensors have limited memory, CPU, and storage capabilities, making it difficult (if not impossible) to deploy resource-intensive AI models. The challenge of limited computational resources requires the development of lightweight AI models optimized for edge devices.

Energy Constraints

Another main challenge is that edge and IoT devices are usually battery-powered, which means that energy efficiency is critical. Deploying AI algorithms that need high computational power can rapidly drain the battery of these devices. This required the development of low-power AI models that could operate efficiently within energy constraints. To solve the problem of energy constraints, hardware accelerators like Neural Processing Units (NPUs), Tensor Processing Units (TPUs), and Graphics Processing Units (GPUs) are being developed for edge devices. These processors can perform AI computations more efficiently, extending battery life and reducing energy consumption.

Heterogeneity of Devices

The heterogeneity of IoT and edge devices introduces a challenge for deploying AI solutions. Devices may have various operating systems, hardware architectures, and connectivity protocols. This complicates the task of developing AI models that can be deployed across diverse environments seamlessly. Developing standardized tools and frameworks that support cross-platform compatibility is required to address this challenge. Platforms such as ONNX Runtime and TensorFlow Lite provide capabilities for deploying AI models over a wide range of edge devices. This simplifies the process of adapting models to various hardware configurations.

Security Vulnerabilities

IoT and edge devices are usually physically accessible. This makes them vulnerable to attacks and tampering. In addition, the limited computational resources of these devices make implementing complex security protocols difficult, which raises the risk of exploitation by adversaries. To improve the security at the edge, techniques like secure boot, device authentication, and hardware-based encryption are used. These techniques ensure that data is encrypted both in transit and at rest and only trusted software can run on edge devices. This decreases the risk of unauthorized access.

Distributed Learning and Federated Learning

We said that one of the challenges in edge computing is the limited resources at the edge. To overcome this, distributed learning methods have been developed. These methods allow several edge devices to collaboratively train AI models without requiring the sharing of raw data. One of the most important approaches is federated learning.

Federated Learning

Federated learning is a distributed machine learning method that makes training possible on decentralized data on edge devices. In federated learning, instead of transferring raw data to a central server, each device trains its local model using its data; then, only the model updates are sent back to a central aggregator. This approach provides some key benefits, such as below:

- **Privacy**: Federated learning improves privacy as data never leaves the edge device. This decreases the risk of data breaches.
- **Reduced Bandwidth**: By transmitting only model updates, not raw data, federated learning decreases bandwidth consumption.
- **Scalability**: Another benefit of federated learning is that it is highly scalable. Several devices can contribute to model training in parallel, which allows for large-scale collaborative learning.

Federated learning is especially helpful in applications where data privacy is a top priority. These cases include healthcare and finance. For instance, hospitals can collaboratively train a model to diagnose diseases based on patient data without sharing sensitive medical data between institutions.

Before continuing our discussion about the challenges of federated learning, let's have some sample code to see how it works. The code below shows a simple federated learning.

Listing 10-5. Simple Federated Learning

```
#Federated Learning
import random

class Server:
    def __init__(self):
        self.global_model = None  # Initialize later

    def aggregate_updates(self, client_updates):
        # Simple averaging for demonstration (replace with more
        sophisticated aggregation)
        avg_weights = {}
```

```
        for key in client_updates[0].keys():            avg_weights[key] =
        sum(update[key] for update in client_updates) / len(client_updates)
        self.global_model = avg_weights  # Update the global model

class Client:
    def __init__(self, data):
        self.data = data
        self.local_model = {} # Initialize a model (a simple dictionary
        for now)

    def train_local_model(self):
        # Simulate local training (replace with actual training)
        for feature in self.data:
            self.local_model[feature] = random.random() # Assign random
            weights for demonstration

    def send_update(self, server):
        return self.local_model  # Send the updated model weights

# Example usage
server = Server()
server.global_model = {"feature1": 0.5, "feature2": 0.2, "feature3": 0.8} #
Initialize the global model

# Sample Data (replace with your actual data)
data1 = ["feature1", "feature2"]
data2 = ["feature2", "feature3"]
data3 = ["feature1", "feature3"]

clients = [Client(data1), Client(data2), Client(data3)]

for _ in range(3):  # Simulate a few rounds of training
    client_updates = []
    for client in clients:
        client.local_model = server.global_model.copy() # Start training
        with the global model
        client.train_local_model()
        update = client.send_update(server)
        client_updates.append(update)
```

```
    server.aggregate_updates(client_updates)
    print(f"Global Model (Round {_+1}): {server.global_model}")
```

Expected Results

```
Global Model (Round 1): {'feature1': 0.5276299384909172, 'feature2':
0.31767992528471845, 'feature3': 0.6518482824530992}
Global Model (Round 2): {'feature1': 0.5138682585770584, 'feature2':
0.6317400687892429, 'feature3': 0.4877414440360856}
Global Model (Round 3): {'feature1': 0.3147066614051861, 'feature2':
0.5343003780938445, 'feature3': 0.5358006602571966}
```

This example illustrates the core principle of federated learning. In this, local model updates are combined to form an overall robust model without compromising individual data privacy.

Challenges in Federated Learning

Although federated learning has significant advantages, it also has some challenges as follows:

- **Communication Overhead**: Frequent communication between the central aggregator and edge devices leads to high communication overhead. This is problematic, especially in environments with limited network bandwidth.
- **Non-IID Data**: Data on edge devices is usually non-independent and identically distributed (non-IID). It means that data distributions can vary considerably across devices. This heterogeneity is a big challenge for training a global model that generalizes well across all edge devices.
- **Privacy and Security**: While federated learning improves privacy by keeping data locally, it is still vulnerable to attacks, like membership inference and model inversion attacks. Secure aggregation and differential privacy techniques are used to address these vulnerabilities and ensure that individual contributions remain confidential.

Edge Inference

The deployment of AI models on edge devices for real-time decision-making is referred to as edge inference. Despite training, which is resource-intensive, in inference, pre-trained models are used to make predictions on new data.

Lightweight AI Models for Edge Inference

Because of the limited resources at the edge, deploying lightweight AI models is necessary for effective edge inference. Some best practices include

Model Compression: Techniques like pruning, quantization, and knowledge distillation are good techniques for reducing the size of AI models and making them suitable for edge devices.

Quantization: This technique converts model parameters from high-precision floating-point to lower-precision formats, like 8-bit integers. This technique reduces both computation and memory requirements.

Pruning: Pruning removes redundant connections and parameters in the model, resulting in a more efficient, smaller model.

Knowledge Distillation: In this approach, a smaller "student" model is trained to replicate the behavior of a larger "teacher" model. This approach achieves similar performance with less complexity.

TinyML: TinyMLs are machine learning models particularly designed for ultra-low-power devices. These models are optimized to run on resource-constrained environments such as microcontrollers.

TinyML has made it possible to deploy AI models on devices with extremely limited power, like wearables. This enables implementing machine learning in applications where power efficiency is important.

Use Cases of Edge Inference

In this part, we discuss some use cases of edge devices. One common edge device is smart cameras. Smart surveillance cameras use edge inference for recognizing faces, detecting anomalies, and tracking objects in real time without relying on cloud processing. This reduces bandwidth usage and makes rapid decision-making possible by only transmitting relevant events to the cloud. Another sample is wearable health devices. These devices use edge inference to detect abnormalities, monitor vital signs, and alert healthcare providers or users when necessary. For instance, a smartwatch can detect irregular heartbeats and can alert the user rapidly without cloud dependency. Also, in manufacturing, edge inference can help in predictive maintenance. In these cases, sensor data are analyzed to predict equipment failures before they happen. This decreases maintenance costs and downtime and improves operational efficiency.

Security and Privacy in Edge AI

Despite many benefits that edge AI offers, it has some vulnerabilities, too. In this section, we discuss security and privacy issues related to edge AI.

Security Threats

IoT and edge devices are vulnerable to a wide range of security threats, such as data interception, physical tampering, and remote exploitation. Because of the distributed nature of edge environments, ensuring security is a complicated task that needs addressing threats at several levels, from communication protocols to device security. Here, we discuss two main categories of attacks. The first one is physical attacks. Edge devices are usually deployed in environments where they can be physically tampered or accessed. Attackers can exploit this physical accessibility to extract sensitive information or modify device firmware. Another type of attack is a network attack. As edge devices communicate over networks, they are susceptible to attacks such as denial of service (DoS), man-in-the-middle (MITM), and eavesdropping. Securing communication channels through secure protocols and encryption is necessary to mitigate these threats.

Privacy Preservation

Privacy-preserving methods are necessary for protecting user data in edge AI applications. In addition to federated learning, other methods like secure multiparty computation and homomorphic encryption can be used to perform computations on encrypted data. This approach ensures that sensitive information is secure during processing. Homomorphic encryption performs computations directly on encrypted data without decrypting it. This keeps that data secure throughout the computation process and provides strong privacy guarantees. Secure multiparty computation (SMPC) enables several parties to jointly compute a function over their inputs while maintaining those inputs private. This is especially helpful in scenarios where data from several edge devices needs to be aggregated without revealing individual data points.

Blockchain for Edge Security

Blockchain technology can be used to improve the security of edge AI systems. Blockchain's distributed ledger makes a tamper-proof record of all transactions and can be used for verifying data integrity, managing device identities, and establishing trust between edge devices. This decentralized approach is especially useful in large-scale IoT networks in which centralized security management is impossible.

For instance, blockchain can be used in smart grids to manage and authenticate energy transactions between edge devices. This approach ensures that all transactions are verifiable and secure. Also, the decentralized nature of blockchain makes it robust to single points of failure.

Applications of AI in Edge Computing and IoT

We talked a lot about AI in edge computing, and we mentioned some examples, such as wearable devices. In this section, we provide more examples of AI use in IoT and edge devices.

Healthcare and Wearable Devices

Wearable devices like fitness trackers and smartwatches use edge AI to monitor health metrics, like blood oxygen levels, heart rate, and sleep patterns. Edge AI helps these devices provide real-time insights and alerts, helping users to take proactive actions for improving their health. Below, we review some use cases.

Fall Detection: Smartwatches with built-in gyroscopes and accelerometers use edge AI to detect falls. In the case that a fall is detected, the device can send an alert to healthcare providers or emergency contacts, making a quick response possible.

Chronic Disease Management: Wearable devices help monitor patients with chronic conditions like hypertension or diabetes. By processing data locally, these devices alert patients and provide personalized recommendations when their vital signs deviate from the normal range.

Remote Patient Monitoring: Edge AI can make the remote monitoring of patients easy by analyzing data from medical sensors and sending alerts if it sees abnormalities. This is useful for those with limited mobility or elderly patients, as it lets healthcare providers monitor the health of patients without frequent in-person visits.

Industrial IoT (IIoT)

In industrial environments, edge AI is used in quality control, predictive maintenance, and automation. Industrial systems can analyze sensor data at the edge and predict failures, detect anomalies, and optimize operations without relying on cloud infrastructure. This approach reduces operational costs, improves reliability, and improves overall efficiency.

Predictive Maintenance: Edge AI can analyze sensor data from industrial equipment and predict the time that maintenance is needed. By detecting potential problems before they occur and lead to equipment failure, companies can reduce downtime and decrease maintenance costs.

Quality Control: In industry, edge AI can be used to do product inspection in real time on the production line. Cameras equipped with AI models can detect deviations from quality standards or defects, which allows for immediate corrective action.

Process Optimization: Edge AI can analyze sensor data and adjust parameters in real time to optimize industrial processes. For instance, in a chemical plant, edge AI can monitor pressure, temperature, and other variables to ensure optimal production conditions.

Retail and Smart Homes

AI at the edge is transforming the smart home environments and retail industry by enabling real-time automation, personalization, and improved customer experiences.

Retail: In retail, edge AI can analyze customer behavior, manage inventory, and market in a personalized way. AI can track inventory levels in real time using smart shelves equipped with edge. Also, cameras can analyze customer behavior to provide insights for promotions and targeted advertising.

Smart Homes: Edge AI is making homes smarter by enabling smart thermostats, voice assistants, and security systems. For example, smart speakers use edge AI for natural language processing and voice recognition. This allows users to control home devices without relying on cloud servers. Also, smart doorbells that are equipped with edge AI can recognize faces and alert homeowners of any suspicious activity.

Agriculture

AI at the edge is improving agriculture by enabling automation and precision farming. Edge devices like sensors, drones, and robots can optimize resource usage, monitor crops, and improve productivity.

Precision Farming: Edge AI makes it possible for farmers to make data-driven decisions. They make these decisions by analyzing data from weather stations, soil sensors, and drones. For instance, soil moisture sensors equipped with edge AI can give real-time information about soil conditions and soil humidity, which reduces water consumption.

Crop Monitoring: Drones equipped with edge AI can fly over products and capture images of fields. Then, they analyze the images to detect pest infestation, signs of disease, or nutrient deficiencies. By local data processing, these drones can provide rapid information, which allows farmers to act rapidly.

Livestock Monitoring: There are wearable devices for livestock that use edge AI for monitoring animal health and behavior. For example, edge AI can be used to identify early signs of illness in cattle or other animals, allowing farmers to provide early medical treatment.

Hardware for Edge AI

Deploying AI models on edge devices needs specific hardware that can support the computational needs of edge inference. This hardware should be power-efficient, too. Modern hardware technology has made powerful AI capabilities at the edge possible.

Edge AI Accelerators

Edge AI accelerators are hardware components specifically designed for performing efficient AI computations. These accelerators include Field-Programmable Gate Arrays (FPGAs), Neural Processing Units (NPUs), Tensor Processing Units (TPUs), and Application-Specific Integrated Circuits (ASICs). In the following, we provide a short description for each.

Tensor Processing Units (TPUs): TPUs are designed specifically for high-speed matrix computations. Matrix computations are very common in AI workloads. TPUs can be used in edge devices for accelerating deep learning inference with low power consumption.

Neural Processing Units (NPUs): NPUs are specialized processors that are optimized for neural network operation executions. They are usually integrated into IoT devices and mobiles to enable efficient AI inference.

Field-Programmable Gate Arrays (FPGAs): FPGAs provide a flexible hardware platform for AI acceleration. FPGAs can optimize specific AI models, which makes them suitable for edge devices.

Application-Specific Integrated Circuits (ASICs): ASICs are custom-designed chips and can be optimized for some specific tasks. In edge AI, ASICs can execute AI models with low power usage and high efficiency. An example of an ASIC designed for edge AI applications is Google's Edge TPU.

Embedded Systems

Embedded systems play an important role in edge AI. Embedded systems are the computing units within IoT devices. These systems are optimized for running specific AI tasks, including image recognition, sensor data processing, and signal analysis.

Microcontrollers: Microcontrollers are widely used in edge AI because of their small form factor and low power consumption. They are perfect for running lightweight AI models for tasks like signal processing and anomaly detection.

Single-Board Computers (SBCs): SBCs, like the NVIDIA Jetson Nano and Raspberry Pi, are commonly used for deploying edge AI applications and prototyping. These boards have enough computational power for running AI models and are suitable for applications in smart cameras, robotics, and industrial automation.

Summary

In the growing area of machine learning, realizing future trends and their potential effect is the main topic for maintaining a competitive advantage. Trends such as transformer architectures, federated learning from an algorithmic point of view, and edge computing from an architectural perspective are the future of ML. In this chapter, we talked about how organizations must take care of these concerns. Additionally, we provided information about how cutting-edge technologies are changing ML. Now, we can finish the chapter and also the book. It was a very good time for me to talk with you about various features of machine learning architecture, but unfortunately, we must finish our discussion here. I hope you have learned interesting topics from this book.

Bibliography

1. Taye, M. M. Understanding of machine learning with deep learning: architectures, workflow, applications and future directions. *Computers* **12**, 91–91 (2023).

2. Nazir, R., Bucaioni, A. & Pelliccione, P. Architecting ML-enabled systems: Challenges, best practices, and design decisions. *J. Syst. Softw.* **207**, 111860–111860 (2024).

3. Beam, A. L., Manrai, A. K. & Ghassemi, M. Challenges to the reproducibility of machine learning models in health care. *Jama* **323**, 305–306 (2020).

4. Ashmore, R., Calinescu, R. & Paterson, C. Assuring the machine learning lifecycle: Desiderata, methods, and challenges. *ACM Comput. Surv. CSUR* **54**, 1–39 (2021).

5. Whang, S. E., Roh, Y., Song, H. & Lee, J.-G. Data collection and quality challenges in deep learning: A data-centric AI perspective. *VLDB J.* **32**, 791–813 (2023).

6. Mahdiani, M. R. Enhancing Recommendation Systems with Federated Learning for Privacy-Preserving E-Commerce Platforms. *Available SSRN 5129722.*

7. Verdonck, T., Baesens, B., Óskarsdóttir, M. & vanden Broucke, S. Special issue on feature engineering editorial. *Mach. Learn.* **113**, 3917–3928 (2024).

8. Mahdiani, M. R. & Evangelista, J. Convolutional Neural Network for Dyno Card Classification: A Deep Learning Approach to Predictive Maintenance in Industrial Systems. *Available SSRN 4633288.*

M. R. Mahdiani, *Mastering Machine Learning Architecture and Solutions*,
https://doi.org/10.1007/979-8-8688-2527-9

9. Mahdiani, M. R. & Khamehchi, E. A modified neural network model for predicting the crude oil price. *Intellect. Econ.* **10**, 71–77 (2016).

10. Mahdiani, M. R., Khamehchı, E. & Suratgar, A. A. A new time-dependent model for controlling the gas injection pressure in continuous gas lift. *Sigma J. Eng. Nat. Sci.* **38**, 265–279 (2020).

11. Mahdiani, M. R., Khamehchi, E., Hajirezaie, S. & Hemmati-Sarapardeh, A. Modeling viscosity of crude oil using k-nearest neighbor algorithm. *Adv. Geo-Energy Res.* **4**, 435–447 (2020).

12. Zhu, B. & Liu, Y. General approximate cross validation for model selection: Supervised, semi-supervised and pairwise learning. in 5281–5289 (2021).

13. Mahdiani, M. R., Khamehchi, E., Soltanmohammadi, R. & Azkayi, B. A New Proxy Model , Based On Meta heuristic Algorithms For Estimating Gas Compressor Torque. in (Tehran, 2015).

14. Yang, L. & Shami, A. On hyperparameter optimization of machine learning algorithms: Theory and practice. *Neurocomputing* **415**, 295–316 (2020).

15. Raiaan, M. A. K. *et al.* A systematic review of hyperparameter optimization techniques in Convolutional Neural Networks. *Decis. Anal. J.* 100470–100470 (2024).

16. Mahdiani, M. R. & Khamehchi, E. A New Method for Building Proxy Models Using Simulated Annealing. **22**, 324–328 (2014).

17. Khamehchi, E. & Mahdiani, M. R. Optimization Algorithms. in *Gas Allocation Optimization Methods in Artificial Gas Lift* 35–46 (Springer International Publishing, 2017).

18. Khamehchi, E. & Mahdiani, M. R. Constraint Optimization. in *Gas Allocation Optimization Methods in Artificial Gas Lift* 25–34 (Springer International Publishing, 2017).

19. Prasetyaningrum, P. T., Purwanto, P. & Rochim, A. F. Enhancing Element Game Classification: Effective Techniques for Handling Imbalanced Classes. *Int. J. Intell. Eng. Syst.* **17**, (2024).

20. Mohammad Reza Mahdiani. *Building .Net Application: C# Programming*. (Behavaran, Kalkzarin, 2010).

21. Mohammad Reza Mahdiani & Mohammad Norouzi. *MATLAB Programming in Engineering*. 272 (Setayesh Press, 2016).

22. Ataei, P. & Staegemann, D. Application of microservices patterns to big data systems. *J. Big Data* **10**, 56–56 (2023).

23. Wang, Y., Kadiyala, H. & Rubin, J. Promises and challenges of microservices: an exploratory study. *Empir. Softw. Eng.* **26**, 63–63 (2021).

24. Lwakatare, L. E., Raj, A., Crnkovic, I., Bosch, J. & Olsson, H. H. Large-scale machine learning systems in real-world industrial settings: A review of challenges and solutions. *Inf. Softw. Technol.* **127**, 106368–106368 (2020).

25. Fortuna, C. *et al.* On-premise artificial intelligence as a service for small and medium size setups. in *Advances in Engineering and Information Science Toward Smart City and Beyond* 53–73 (Springer, 2023).

26. Kumar, S. *et al.* Applications, challenges, and future directions of human-in-the-loop learning. *IEEE Access* (2024).

27. Chahal, D., Mishra, M., Palepu, S. & Singhal, R. Performance and cost comparison of cloud services for deep learning workload. in 49–55 (2021).

28. Wang, Y. E., Wei, G.-Y. & Brooks, D. Benchmarking TPU, GPU, and CPU platforms for deep learning. *ArXiv Prepr. ArXiv190710701* (2019).

29. Alser, M. *et al.* Packaging and containerization of computational methods. *Nat. Protoc.* **19**, 2529–2539 (2024).

30. Seo, J., Lee, Y. & Yoon, Y. Self-sovereign and Secure Data Sharing Through Docker Containers for Machine Learning on Remote Node. *J. Web Eng.* **23**, 637–655 (2024).

31. Egbuna, O. P. Machine Learning Applications in Kubernetes for Autonomous Container Management. *J. Artif. Intell. Res.* **4**, 196–219 (2024).

32. Casalicchio, E. & Iannucci, S. The state-of-the-art in container technologies: Application, orchestration and security. *Concurr. Comput. Pract. Exp.* **32**, e5668–e5668 (2020).

33. Nasir, V. & Sassani, F. A review on deep learning in machining and tool monitoring: Methods, opportunities, and challenges. *Int. J. Adv. Manuf. Technol.* **115**, 2683–2709 (2021).

34. Retzlaff, C. O. *et al.* Human-in-the-loop reinforcement learning: A survey and position on requirements, challenges, and opportunities. *J. Artif. Intell. Res.* **79**, 359–415 (2024).

35. Wu, Y., Dobriban, E. & Davidson, S. Deltagrad: Rapid retraining of machine learning models. in 10355–10366 (PMLR, 2020).

36. Saha, G. C. *et al.* Human-AI collaboration: Exploring interfaces for interactive machine learning. *Tuijin JishuJournal Propuls. Technol.* **44**, 2023-2023 (2023).

37. Emami, Y., Almeida, L., Li, K., Ni, W. & Han, Z. Human-in-the-loop machine learning for safe and ethical autonomous vehicles: Principles, challenges, and opportunities. *ArXiv Prepr. ArXiv240812548* (2024).

38. Pulicharla, M. R. Data Versioning and Its Impact on Machine Learning Models. *J. Sci. Technol.* **5**, 22–37 (2024).

39. Spadari, V., Cerasuolo, F., Bovenzi, G. & Pescapè, A. An MLOps Framework for Explainable Network Intrusion Detection with MLflow. in 1–6 (IEEE, 2024).

40. Jiang, W. *et al.* Challenges and practices of deep learning model reengineering: A case study on computer vision. *Empir. Softw. Eng.* **29**, 142–142 (2024).

41. Wang, Y. *et al.* Deep adversarial domain adaptation for breast cancer screening from mammograms. *Med. Image Anal.* **73**, 102147–102147 (2021).

42. Mahdiani, M. R. & Khamehchi, E. Using mapping for increasing the speed and quality of meta heuristic optimization (Case Study: Petroleum Engineering). in (2015).

43. Grafberger, S., Groth, P., Stoyanovich, J. & Schelter, S. Data distribution debugging in machine learning pipelines. *VLDB J.* **31**, 1103–1126 (2022).

44. Agrahari, S. & Singh, A. K. Concept drift detection in data stream mining: A literature review. *J. King Saud Univ.-Comput. Inf. Sci.* **34**, 9523–9540 (2022).

45. Bejani, M. M. & Ghatee, M. A systematic review on overfitting control in shallow and deep neural networks. *Artif. Intell. Rev.* **54**, 6391–6438 (2021).

46. Pagano, T. P. *et al.* Bias and Unfairness in Machine Learning Models: A Systematic Review on Datasets, Tools, Fairness Metrics, and Identification and Mitigation Methods. Big Data and Cognitive Computing, 7, 15. (2023).

47. Koppisetti, V. S. K. The Role of Explainable AI in Building Trustworthy Machine Learning Systems. *ESP Int. J. Adv. Sci. Technol. ESPIJAST Vol.* **2**, 16–21 (2024).

48. Rawal, A., McCoy, J., Rawat, D. B., Sadler, B. M. & Amant, R. S. Recent advances in trustworthy explainable artificial intelligence: Status, challenges, and perspectives. *IEEE Trans. Artif. Intell.* **3**, 852–866 (2021).

49. Zhu, X., Hu, X., Yang, L., Pedrycz, W. & Li, Z. A development of fuzzy-rule-based regression models through using decision trees. *IEEE Trans. Fuzzy Syst.* **32**, 2976–2986 (2024).

50. Esterhuizen, J. A., Goldsmith, B. R. & Linic, S. Interpretable machine learning for knowledge generation in heterogeneous catalysis. *Nat. Catal.* **5**, 175–184 (2022).

51. Chockler, H., Kelly, D. A., Kroening, D. & Sun, Y. Causal Explanations for Image Classifiers. *ArXiv Prepr. ArXiv241108875* (2024).

52. Mendonça, M. O. K., Netto, S. L., Diniz, P. S. R. & Theodoridis, S. Machine learning: Review and trends. *Signal Process. Mach. Learn. Theory* 869–959 (2024).

53. Gillioz, A., Casas, J., Mugellini, E. & Abou Khaled, O. Overview of the Transformer-based Models for NLP Tasks. in 179–183 (IEEE, 2020).

54. Thisanke, H. *et al.* Semantic segmentation using Vision Transformers: A survey. *Eng. Appl. Artif. Intell.* **126**, 106669–106669 (2023).

Index

A

M. R. Mahdiani, *Mastering Machine Learning Architecture and Solutions*,
https://doi.org/10.1007/979-8-8688-2527-9

E

F

G

N

Q

R

S

T

U

V

W

X, Y

Z